L O F T S

LOFTS 改造与更新设计
——城市居住的新设想

[美] 费莉西娅·艾森伯格·莫尔纳 著

重庆仁豪景观设计有限公司 译

中国建筑工业出版社

著作权合同登记图字：01-2002-2257号

图书在版编目（CIP）数据

LOFTS改造与更新设计——城市居住的新设想／（美）莫尔纳著．重庆仁豪景观设计有限公司译．—北京：中国建筑工业出版社，2003
 ISBN 7-112-05824-4

Ⅰ．LOFTS... Ⅱ．①莫... ②重... Ⅲ．住宅-建筑设计 Ⅳ．TU241

中国版本图书馆CIP数据核字（2003）第033570号

Copyright © 1999 by Rockport Publishers, Inc
Lofts/Felicia Eisenbarg Molnar
本书由美国Rockport出版社授权翻译出版

责任编辑：程素荣

LOFTS改造与更新设计
——城市居住的新设想

［美］费莉西娅·艾森伯格·莫尔纳 著
 重庆仁豪景观设计有限公司 译

中国建筑工业出版社出版、发行（北京西郊百万庄）
新 华 书 店 经 销
北京嘉泰利德公司制版
利丰雅高印刷（深圳）有限公司印刷
开本：889×1194毫米 1/16 印张：10 字数：250千字
2003年10月第一版 2003年10月第一次印刷
定价：**95.00**元
ISBN 7-112-05824-4
TU·5119（11463）
版权所有 翻印必究
如有印装质量问题，可寄本社退换
（邮政编码100037）
本社网址：http://www.china-abp.com.cn
网上书店：http://www.china-building.com.cn

To Imre and Isabelle and the building of dreams.

献给艾米尔和伊莎贝尔，以及追逐梦想的人们

CONTENTS
目　录

前言	8
阁楼空间：新型的城市居住方式 乔治·拉纳利	10
纽约市的阁楼设计	12
城市边缘阁楼，迪安/沃尔夫建筑师事务所	14
终极阁楼，哈迪·霍尔兹曼·法伊弗联合事务所	18
第34大街阁楼，LOT/EK建筑事务所	24
匡特阁楼，托德·威廉斯与比利·陈联合事务所	28
家庭起居阁楼，斯科特·马波及卡伦·费尔班克斯事务所	32
特里贝卡阁楼，陶工作室	38
雷瑙德阁楼，查&茵纳霍夫尔建筑及设计事务所	42
第五大街阁楼，斯科特·马波及卡伦·费尔班克斯事务所	46
戴夫尼住宅，哈特·萨克斯工作室	50
哈德森阁楼，亚历山大·戈林建筑师事务所	54
魏家阁楼，巴克建筑事务所	58
城市阁楼，维森特·沃尔夫联合公司	62
切尔西阁楼，卡尔·霍建筑师事务所	66
罗森伯格住宅与工作室，贝尔蒙特·弗里曼建筑师事务所	72
生活/工作两用阁楼，迪安/沃尔夫建筑师事务所	76
艺术家工作室，加里亚·索罗莫诺夫设计室	80
特里贝卡住宅与工作室，莫内欧·布洛克建筑师事务所	84

走向世界	88
卡尔森—里杰斯住宅，洛托建筑师事务所 洛杉矶	90
混凝土阁楼，弗兰克尔+科尔曼建筑师事务所 芝加哥	94
库伦住宅，克里斯·迈斯建筑师事务所 安特卫普	100
奥立弗码头，麦克道尔+贝内迪提事务所 伦　敦	104
奥马利住宅，卡彭特/格罗津斯建筑师事务所 宾夕法尼亚州，阿沃卡	110
新河源住宅，麦克道尔+贝内迪提事务所 伦　敦	114
帕沃尼亚阁楼，安德斯联合事务所 泽西城	118
尼尔码头住宅，里克·马瑟建筑师事务所 伦　敦	124
奇斯维克·格林工作室，彼得·华德利建筑师事务所 伦　敦	130
特里贝卡阁楼，特里朗奇设计事务所 多伦多	134
卡萨之家图书馆，罗桑那·蒙奇尼事务所 米　兰	138
玛特商品仓库，赛可尼·西蒙事务所 多伦多	142
西新宿住宅，费德里克·费歇尔联合事务所 东　京	146
海湾顶楼仓库，布莱顿与休斯设计工作室 旧金山	150
事务所目录	154
摄影工作室目录	156
照片来源	158

PREFACE
前　言

阁楼[1]住居生活方式的出现常常可以追溯到20世纪50年代末期的曼哈顿小型办公和家庭办公区（SoHo District）。在当时，很难找得到适合于富有创造精神的城市年轻人的工作与生活空间。因此，艺术家和手工艺家们把目光投向了那些被废弃的厂房，厂房巨大的尺度提供了在价格上较为便宜的集工作、居住于一体的工作室，尽管这通常是不合法的。随着城政当局开始意识到这种改造所带来的好处，它们对法律进行了修订，以鼓励二次开发。不久，那些寻求一种对传统城市公寓的替代物的年轻专业人士也加入艺术家的队伍。时尚咖啡店、餐馆和画廊在这个曾经一度被人们所遗忘的街区扎下了根，而这些地方的房地产价值也一路飙升。

在阁楼中居住的方式所具有的魅力是显而易见的。阁楼是浪漫的场所，成排的窗户和高耸的顶棚形成了不寻常的高度、良好的采光效果，以及开放的空间，并让人们得以在拥挤的城市环境中寻找到个性的自由。每一个阁楼都为我们提供了一片广阔的天地，可以让城市居民在这个似乎无止境的空间中任意挥洒。

阁楼虽然充满了令人激动的可能性，但知道如何在其中居住并应付改造所面临的困难却让人望而生畏。阁楼空间少有墙壁，也没有走道，我们该怎么办？如何布置喷淋系统？该如何处理那些让人看不到外面或射进太多阳光的窗户、沾满油漆斑点的混凝土楼面，以及头顶隆隆作响的空调与采暖管道？怎样处理没有划分成卧室、浴室和厨房的开敞空间？对于那些没有经过专业训练的人在休量巨大的旧发电厂或冰激淋厂房面前不知所措。在他们看来，一个具有这种无法想像的庞然大物般的阁楼似乎是不适合人类居住的。

但是，通过对各种设计方案的理解，我们已经开始能够领略到大空间的趣味并强化开放性。非个性化逐渐消逝，而梦想从此开始实现。只有巨大的创造力和充分的想像力才能把制革厂或原来的印刷厂改造成一个可以被称为"家"的场所。

---

[1] 阁楼（loft）一词在本书中指位于工厂、仓库或其他商业或工业建筑物上层的巨大而通常未间隔开的楼层，或是在建筑屋顶下的大空间。为统一起见，本书中均翻译为阁楼或阁楼空间。——译注

本书的目的就是要展示这样的一个梦想。它可以激发灵感——简约的现代主义线条确实可以装饰陈旧的风景。本书的宗旨在于揭示建筑师、室内设计师和阁楼居住者是怎样把单调的废弃工业空间编织为一个真正意义上的"家"。

本项目起源于多年来对世界各地城市阁楼的关注，不管是白天还是夜晚，都梦想着设计阁楼，梦想着在阁楼中生活和工作。在一定程度上说，在视觉上享受和品味阁楼还不够；人们还想知道：谁设计了阁楼、谁居住在阁楼中、他们在做什么？

阁楼居住者也许是所有城市人中最引人注目的一族，他们是现代城市的英雄，他们努力拯救和保存城市，而不是摧毁城市，他们把过去的巨大遗址转变为未来的伟大居所。许多城市中的那些曾经被人遗忘的地区已经恢复成这个星球上最激动人心的居住地。

如今，纽约的艺术家、设计师以及阁楼居民仍然在很大程度上引领着阁楼设计的潮流。虽然此类建筑正在日益减少，但在世界各地的城市中去寻找更为荒凉的街区的这种原始创新精神依然在延续。本书以独特的视角，阐述了纽约对世界的阁楼设计所产生的微妙与明显的影响。

LOFT SPACE THE NEW URBAN DWELLING

阁楼空间：新型的城市居住方式　　by George Ranalli 乔治·拉纳利

　　阁楼设计有许多方面需要从理论上进行探讨，对诸如家庭组织方式、多个相互关联的加庭，以及在城市尺度上的由家庭所组成的社区等等，都是值得广泛研究的问题。在阁楼设计中，家庭结构不是预先设定好的，公众和私人在所有层面上的相互影响都应进行公开阐述。阁楼住宅设计赋予了家居生活一种新概念；反过来，这种新型家居概念为全世界的传统住宅设计带来了极大的改变。

　　阁楼住宅最初是一种美国现象，但逐渐演变成全世界许多城市人梦寐以求的居住方式。大多数情况下，那些废弃后被重新发现的建筑所具有的空间又窄又长，窗户在建筑两端或者沿着某一个长向的墙壁一溜排开。不管是何种布局，这些原有条件通常展现在我们面前的是一种强烈的初始状态，并且通常是一览无余的骨架式的结构。

　　设计阁楼空间的挑战在于：要提供必备的舒适的居住（有时候还有工作）环境，但又不能牺牲原来空间的开放性、良好的采光和自由流动性等方面的品质。每一个阁楼都具有一个全新的、不同的、充满挑战性的空间难题。现有的窗户、电梯和楼梯是在阁楼设计的组织中的决定性要素，而这些既定要素通常使手头上的难题具有谜一般的特点。阁楼设计的最终目标是：满足所有必要的采光和空间要求，同时创造出一种与原来建筑保持着某种联系的既吸引人的注意力又具有功能性的设计。

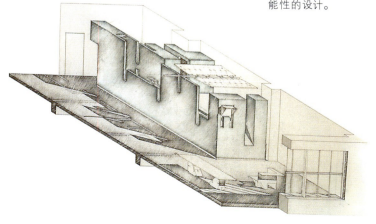

左图：乔治·拉纳利设计的K－阁楼轴测图。该项目为一个2,100平方英尺（195.1m²）的阁楼，将从前工业建筑中的原有砖墙与一系列的新形式和定制装饰构件结合起来

LEFT Axonometric drawing of the K-Loft designed by George Ranalli. The project was a 2,100-square-foot (195.1-square-meter) loft that integrated original brick walls of the former industrial building with a series of new forms and custom decorative objects.

阁楼空间通常拥有丰富的材料，但表面的处理有些粗糙。由于大多数阁楼都曾经是工业空间，那些未经修饰的砖石砌体给环境带来了一种强烈的纹理感，同时为设计师和居住者提供了具有强烈个性和历史特征的空间。

在我的阁楼设计作品中，我始终致力于探索公众与个人家庭生活的需要，试图从中产生出一些未经尝试的新型家居安排的可能性。我们一直抱着这样一种观念：公共领域是家庭和公众进行庆典活动的空间，而家庭生活的私密空间应该能够让人们居住在他们的思想、感情和梦想之中。对完全开放的阁楼空间所进行的具体改造和诠释能够产生出某些有趣的对空间及家庭的安排和相互关系，而这些是在我们习以为常的居住建筑中所不具备的。比如，本书第118页中的彼特·安德斯阁楼（Peter Anders Loft）就是对两个家庭分享同一个阁楼空间的新型居住方式的研究和尝试。这种居住方式的实现需要新颖的处理手法，包括设置共用区域和在私密空间之间形成过渡的联系要素。

在阁楼设计中，对材料和形式的探索也是一个非常动态的问题。两个要素之间的亲密对话可以采用非常规的方式来进行试验。细部处理也是在阁楼设计中最后一个和最重要的组成部分。对细部的处理应反映复杂的技术与审美观念，它们主要通过其造型和比例来表现建筑的外部美感。

最后，对阁楼居住方式的新颖设计与作为一个历史产物的建筑现状之间的紧密联系才是至关重要的。我们应该努力探索那种能够在材料的美中表现出来的具有连贯性、清晰性、有时候是创新性的家庭生活的观念。这就是在阁楼设计上所面临的挑战，本书后面介绍的许多设计项目都突出了这一点。

乔治·拉纳利是耶鲁大学的建筑设计教授，在纽约市拥有一个私人建筑事务所。拉纳利先生的作品曾在国际上展出，并且是大都会艺术博物馆的20世纪艺术与设计永久收藏品中的一部分。他也曾经是两部专著的讨论对象，他本人的作品也已经在全世界广泛出版。

上图： 新港住宅（Newport Residence），该项目将一个国家级的具有历史意义的标志性学校建筑改建而成六个阁楼单元。空间正对着一个室内立面，其后布置了更小、更具私密性的房间

ABOVE Newport Residence, conversion of a National Register Historic Landmark schoolhouse into six loft units. The spaces are fronted with an interior façade behind which the smaller, more private rooms are arranged.

LOFT DESIGN IN NEW YORK CITY
纽约市的阁楼设计

纽约以充满活力、不受约束和富于想象的文化氛围赢得了世界的普遍羡慕。在20世纪50年代，曼哈顿的寻梦者和幻想家现代社会提供了在改造废弃工业建筑和为一个新的城市居民阶层创造空间提供了一种充满灵感的模式。这个潮流主要是由艺术家和手工艺者率先发起，具有创新性意义，但最初并不合法。市政当局最终意识到这是在城市再开发中的绝佳机会，于是修订法律，鼓励人们改造这些旧的商业区。从而，纽约在阁楼居住方式的设计上引起了世界的关注。

如今，对于那些追求完全开放的空间的人们来说，在曼哈顿的那些被人遗忘的社区中的废弃厂房仍然犹如他们的天堂。在这些空间中存在着令人激动的机会，让他们可以创造性地表现自我和实现大尺度的居住方式。在纽约的阁楼居住生活方式中，传统的形式得以被分析和重新组合。从工业化向居室化的转变是一个有机的过程，它形成于周围的环境。陈旧的皮革厂房、银行大楼或食品库房都是一个设计上的完美开端。废弃的桁架及其他工业构件可按非标准的方式进行重新组装。有时还创造出一些几何形状，以通过结构来组织各个工业构件。很多情况中，在混凝土梁或钢梁，通风管道，或是货运电梯等要素上依然保留这些空间所具有的历史痕迹。它们所具有的美感和实用性让人能够感受到与建筑的原有特征之间的紧密联系。

极简主义的理想和功能性的需要成为决定纽约最早一批居住和工作两用阁楼的室内设计的因素。在这样一个"少就是多"的时代里，复杂已成为过去，人们依然注重着极简主义的理想。然而，本书后面所展示出来的丰富的材料和精美工艺反驳了关于这个愚人村[1]（Gotham City）的设计风格是冰冷的、臣服的和不可触摸的论调。玻璃和钢铁因其宝石般的精密度而熠熠生辉；木材展现出极富质感的纹理；混凝土也可变得充满温情。在纽约的许多阁楼上所具有的这些品质不断地给欧洲及其他地方的设计师带来灵感。曼哈顿的设计风格告诉我们：即使空间非常珍贵，但阁楼式的住宅也可显得宽绰。在阁楼的设计过程中，纽约人渴望建立一种寄托个人梦想的"乌托邦"空间——而这似乎是一个全世界都想追逐和建立的潮流。

---

[1] Gotham City 是纽约市的别称。——译者注

Urban Interface Loft
DEAN/WOLF ARCHITECTS

城市边缘阁楼
迪安/沃尔夫建筑师事务所

在许多缺乏自然采光的城市公寓里居住过之后，业主兼设计师试图去寻找一个可以让城市与天空相接的家。在高楼林立的曼哈顿，要想实现这个梦想并不是易事。经过一段长时间的搜寻，这对夫妇和他们的朋友在特里贝卡看上了一整幢六层楼的废弃电气仓库。难以置信的是，有这么一家金融合伙企业愿意来投资这个项目，而凯思林·迪安（Kathryn Dean）和查尔斯·沃尔夫（Charles Wolf）也同意担任该项目的建筑师，对整幢建筑进行全面的改造，但作为交换条件，他们将会得到最具有价值的顶层空间。虽然这对夫妇自己的阁楼空间显得很狭小，但设计以头上的天空为中心，显得非常的壮观。在设计上最重要的处理是将屋顶凿空，从而形成了一个天井。阳光和景色透过天窗和采光口被引向内部空间。在建筑内部，一道喷砂玻璃隔断将工作室与居住区分隔开来。铜制护墙板与其他材料的配合极具匠心。

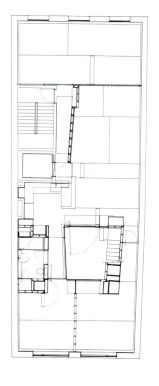

上图：把屋顶凿空形成一个天井，阁楼以天空为中心

左图：顶层平面图

对页图：城市在黄昏时分渐渐隐退，而阁楼成为亮点

ABOVE The loft is focused on the skies above with a courtyard cut out of the roof.

LEFT Top-floor plan.

OPPOSITE At twilight the city recedes and the loft comes into focus.

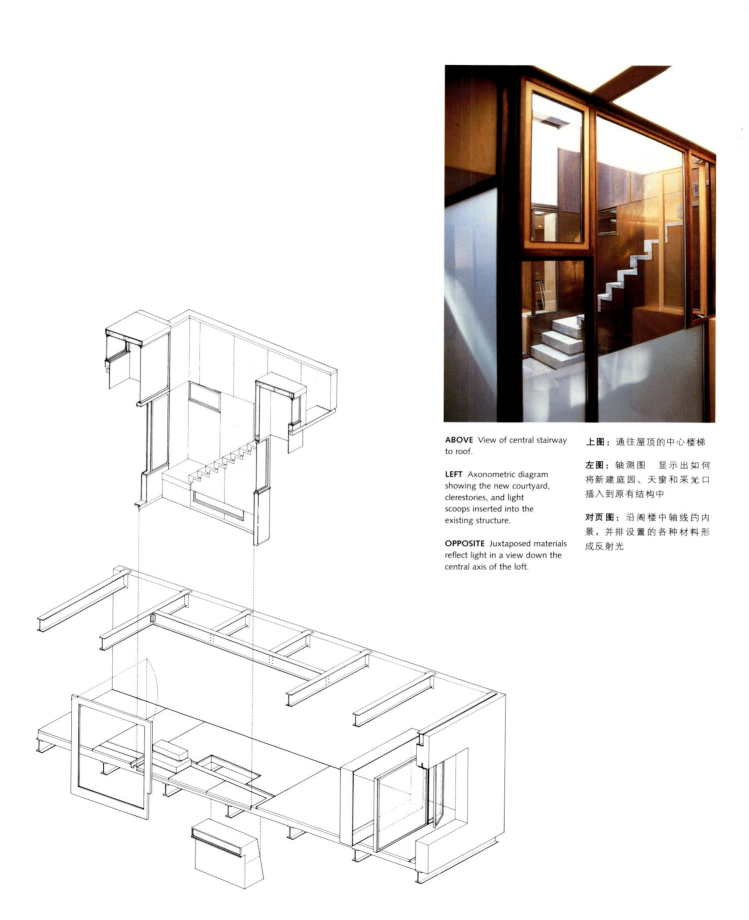

ABOVE View of central stairway to roof.

LEFT Axonometric diagram showing the new courtyard, clerestories, and light scoops inserted into the existing structure.

OPPOSITE Juxtaposed materials reflect light in a view down the central axis of the loft.

上图：通往屋顶的中心楼梯

左图：轴测图 显示出如何将新建庭园、天窗和采光口插入到原有结构中

对页图：沿阁楼中轴线的内景，并排设置的各种材料形成反射光

URBAN INTERFACE LOFT 17

The Ultimate Loft
HARDY HOLZMAN PFEIFFER ASSOCIATES

终极阁楼
哈迪·霍尔兹曼·法伊弗联合事务所

大多数阁楼居住者似乎都倾向于极简主义式的室内装修，拒绝与外界城市的混乱喧嚣有任何雷同之处。但建筑师马尔科姆·霍尔兹曼（Malcolm Holzman）和他的妻子安德烈亚·兰兹曼（Andrea Landsman）却与众不同。他们在自己位于曼哈顿的阁楼里塞满了从外面找来的各种稀奇古怪的杂物，并且还戏称这些收藏品为垃圾。霍尔兹曼说："在我们的阁楼中，我做了一些我的客户不允许我做的事情。"除了夫妇俩收藏的一些著名的美国现实主义绘画和当代雕塑作品之外，这个阁楼是对游戏式的古怪想法所作的尝试。建筑师非常关注不常用的材料及饰面效果。有一些墙上用带错缝式砖和石材图案的电镀压型钢板饰面。其他材料很容易在加油站、高速公路边的带形建筑或工业建筑上找到，包括带绿点的刨花板和波形玻纤瓦。但从整体上看，室内的构图和选色极富技巧性。该阁楼是新观念的试验场，尝试了某些尚未成型的观念。

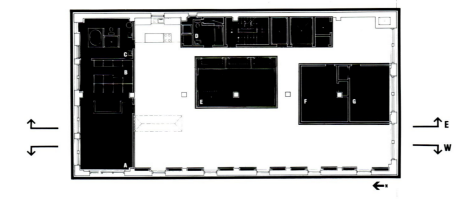

上图：阁楼外的风景。人形的灯具为哈里·安德森（Harry Anderson）的设计

左图：平面图

对页图：材料和颜色在宽敞的居住区里形成鲜明的对比

ABOVE View out of the loft. Anthropomorphic lamp is by Harry Anderson.

LEFT Floor plan.

OPPOSITE Materials and colors collide in the spacious living area.

ABOVE Galvanized, stamped steel in alternating brick and stone patterns on the wall. Dining table consists of slabs of Georgia marble.

OPPOSITE A Blatz Beer sign ornaments the kitchen. The plates are gifts from friends and are accepted by the owners only if they cost less than one U.S. dollar.

上图：带错缝式砖和石材图案的电镀压型钢板用于墙面装饰。餐桌用乔治亚大理石板做成

对页图：厨房用布拉仔牌啤酒标志进行装饰。盘子是朋友们赠送的礼物，主人们会在它们的价格不超过1美元的时候才会接受下来

右图：厨房一景，旁边是半透明的玻璃纤维墙，保留了原有的混凝土梁和柱子

RIGHT A view toward the kitchen and adjacent walls of translucent fiberglass, with original concrete beams and columns.

34th Street Loft
LOT/EK ARCHITECTURE

第34大街阁楼
LOT/EK 建筑事务所

LOT/EK建筑事务所为"拾得的物品"这个名词赋予了新的含义。在这个改造之后的阁楼中，古伊斯皮·李格纳诺（Guisseppe Lugnano）和阿达·托拉（Ada Tolla）展示了他们在建筑设计上高超的魔术般的技艺。作为设计师，他们对维护这个位于曼哈顿时装区的面积为2000平方英尺（合185.8m^2）的阁楼所具有的内在美感和个性非常感兴趣。在卧室、厨房和浴室区实现业主所渴望的开放性和对私密性的要求并非易事。为了解决这个难题，LOT/EK建筑事务所从外面搬来一个边长为40英尺（合12.2m）的铝制集装箱，用它来作为主要的空间组织元素。金属板穿过空间，把生活空间与业主的工作室分隔开来。为了提高空间的灵活性和方便出入，对金属板进行了各种切割，然后装上铰链或转轴以便开启。三个通高的铝板可以旋转，以遮挡住卧室，另外还用了一些较小的安有铰链的面板来挡住厨房设备。这一铝板隔断在盥洗室结束，在这里，它将淋浴间和卫生间与洗手前室分隔开来。

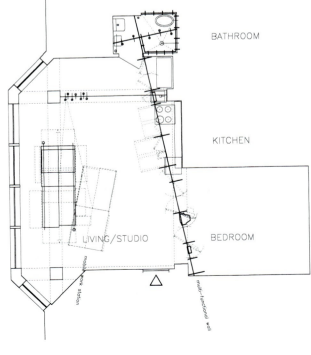

上图及对页图： 铝隔墙上的面板，隔墙贯穿整个阁楼，将厨房设备遮挡住。煤气炉上面的面板像车库门那样向上开启。当把面板关闭起来的时候，只有水槽、烤箱和集装箱的轮廓露在外面。

左图： 平面图

ABOVE AND OPPOSITE Panels in an aluminum plane, which runs the length of the loft, conceal kitchen equipment; the panel above the stove opens up, like a garage door. When closed, only the sink, oven and shipping container graphics are revealed.

LEFT Floor plan.

ABOVE Hinged panel doors cut into the aluminum plane pivot to open up the bedroom to the main space. A television, set into a panel door, is visible in the bedroom when the panels are open and from the larger living area when the doors pivot shut.

RIGHT Recycled refrigerators provide both storage and work surface for the artist couple who occupy this 2,000-square-foot (185.8-square-meter) loft space. Set perpendicular to the aluminum metal plane, a pipe carries bundled electrical cords to provide lighting and power to the work stations.

OPPOSITE The aluminum plane runs the length of the loft, concealing the kitchen, bedroom, and bath from working areas. Existing concrete floors were polished.

上图：切入铝板隔墙的铝板门安装了铰链，打开后将卧室与主空间连为一体。电视机设置在其中一个门扇上，当铝板门打开后，可以在卧室里看电视，而当铝板门关上时，可以在较大的起居区里看电视

右图：回收的旧电冰箱既可用做储藏，也作为居住在这个2000平方英尺（合185.8m²）的阁楼空间的夫妇俩使用的工作台面。一根与铝制金属隔墙相垂直的管子上的成捆的电缆为工作区提供了照明和必需的电源

对页图：铝板隔墙横穿整个阁楼，将厨房区遮挡起来。原有的混凝土楼面经过了重新抛光处理

Quandt Loft
TOD WILLIAMS, BILLIE TSIEN AND ASSOCIATES

匡特阁楼
托德·威廉斯与比利·陈联合事务所

这个非常和谐的阁楼位于纽约市的格林威治村，面积为5000平方英尺（合 464.5m^2），体现出一种冷峻、严肃和非正式的高雅美感。该阁楼由经常能够获奖的夫妻二人组托德·威廉斯和比利·陈进行设计，体现出了和谐、均衡和各个要素之间的张力感。通过采用滑门和活动墙壁（悬挂在顶棚上）来保持宽敞的流动空间。它们围合或开启了在主起居区周围的各种空间。除了在尺寸上较大以外，该阁楼的平面布置非常简单，中间的起居空间处采用水磨石地面，四周是更具有私密性的区域。自然光线从一面大面积开窗的墙的方向射入，同时，彩色抹灰上的各种色调，从紫红色到米黄色，也营造出一种明亮感和动感。建筑师与艺术家、金属件制作商、橱柜厂商紧密协作，对施工中的各个方面都予以了密切的关注。

上图： 一道半透明的墙体依次重叠收缩在轨道上，将盥洗间朝主卧室敞开

左图： 平面图

对页图： 主居住空间内景，中央是用黑色胶合板做成的书架，表现出一种漂浮感

ABOVE A translucent wall telescopes on tracks to open the bathroom to the master bedroom.

LEFT Floor plan.

OPPOSITE Views of the main living space with black plywood bookcases floating in the center.

ABOVE Sleeping and living areas are separated by a sliding wall suspended from an overhead track.

OPPOSITE A cantilevered arm holding a handblown glass flower vase is featured in the entry foyer.

上图：卧室与起居区被从上方滑轨处悬垂下来的一面滑动墙所分隔

对页图：入口大厅里的悬挑式座子固定着一个人工制作的玻璃花瓶

Family Living Loft
SCOTT MARBLE · KAREN FAIRBANKS ARCHITECTS
家庭起居阁楼
斯科特·马波及卡伦·费尔班克斯事务所

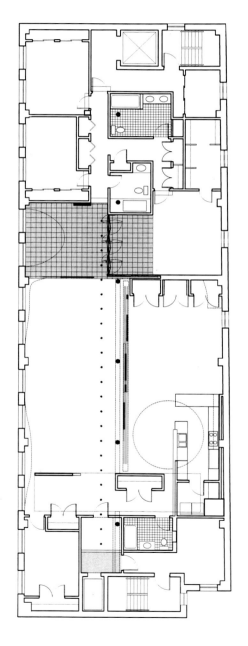

一个四口之家要在拥挤的城市环境中找到灵活的生活空间并非易事。在对这个位于纽约市切尔西地区的面积为 4500 平方英尺（合 418.1m²）的工业建筑改建项目中，设计师为两卧两卫家居式阁楼创造出了一种非传统的家庭组织结构。大胆采用了各种过渡性元素。各种改造要素被充分利用，以协调对私密性的要求和希望获得开放空间的愿望之间的关系。四扇滑动式木玻隔板以及三扇旋转门起着对空间的组织作用，它们将阁楼划分为正式空间和非正式空间。在这个其他方面都非常简单的环境中，在隔板和门扇上的几何形状通过丰富的材料来表现出来，其中包括铝材、木材和刻花玻璃等饰材。由夫妇俩所组成的设计小组将考虑的重点放在阁楼空间适应各种家庭活动的能力上。

LEFT Loft floor plan.

OPPOSITE View from the entry. Sliding panels at right separate formal dining and living areas from informal kitchen and family zones. At the far end, the sliding panel that divides the formal living area from the study is open.

左图：阁楼平面图

对页图：从入口处看室内。右侧的滑动隔板将正式用餐区和起居区与非正式的厨房和家庭生活划分开来。在远端，分隔正式起居区和书房的滑动隔板被打开着

上图：从书房向父母卧室方向看旋转门

ABOVE View of pivoting doors from the study toward parents' bedroom.

对页图：在卧室里铺设的软木地板所具有的暗黑色与其他地方铺设的枫木地板形成强烈的对比

OPPOSITE From within the bedroom cork flooring stained dark lends contrast to maple floors elsewhere.

上图：起居区的内景。在起居区和书房之间的适当位置设置了滑动隔板

对页图：设置在正式和非正式的起居区之间的三块滑动隔板

ABOVE View of living area with sliding panel between living area and study in place.

OPPOSITE View of all three sliding panels fully extended between formal and informal living areas.

Tribeca Loft
TOW STUDIOS

特里贝卡阁楼
陶工作室

该阁楼位于纽约市灯红酒绿的夜总会密集区的中央地带，面积为2800平方英尺（合260.1m²），占据了一幢最近改建过的建筑的整整一层楼。室内空间仅剩下了混凝土梁、楼板和石砌墙体，适合于改建成为一个简单而宁静的、完全开放的两卧两卫的居住空间。建筑的三面墙上开窗，电梯正对着阁楼开门。室内采用了半透明的玻璃门和墙，自然光可以找照到房间的中央部分，甚至还能照到靠里面的更暗的区域。建筑师将枫木和火烧面的法国石灰岩板混合搭配在一起，从而在楼板上形成一种通道式的处理。该通道将西侧的入口和东侧的厨房连接起来。带有滑动玻璃隔板的磨砂玻璃墙南北向设置，像屏风似的将卧室与用餐区分隔开来。当早晨的阳光洒满卧室或者在夜晚时分灯光打开的时候，该屏风就会隐隐发亮。

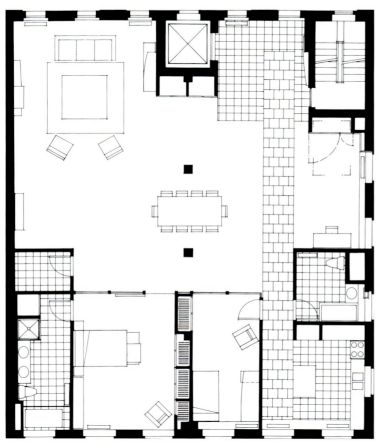

上图： 阁楼平面图

右图： 电梯入口处的墙面用传统的意大利风格的抹灰饰面，在抹灰完成面层里加入了颜料，然后是起沙，最后表面赶光

对页图： 餐厅，还能看到外面的厨房区，地面是用石灰石形成的"小道"和枫木地板

ABOVE Loft floor plan.

RIGHT Elevator entry features stuccoed wall finished in traditional Italian style, adding pigments to the final coat of the plaster and sanding and polishing the surface.

OPPOSITE Dining room with view out to kitchen area, showing limestone "path" and maple floor.

ABOVE The spacious, wide-open main living area.

上图：宽敞开放的主起居区

ABOVE Bedrooms and kitchen are situated along perimeter walls to allow for maximum interior living area.

上图：卧室和厨房沿外墙布置，从而可以获得最大的室内起居区域

Renaud Loft
CHA & INNERHOFFER ARCHITECTURE + DESIGN

雷瑙德阁楼
查&茵纳霍夫尔建筑及设计事务所

雷瑙德阁楼的设计对现代主义关于平面和体量、不透明性与透明性的主题在与古典元素并列放置时所能产生的效果进行了探索。该阁楼原有面积为4000平方英尺（合371.6m^2），位于一个改建过的5层楼高的标志性建筑内，很适合将它改造为供一位年轻的银行家居住的私人住宅。该阁楼被设计成一个远离喧闹城市生活的宁静港湾，可以为款待和收留常客提供一个高雅的环境。设计小组沿两条非常畅通的流线按私密和公共功能将阁楼划分开来。该流线和各个功能之间的相互影响是通过使用各种材料所形成的感官上的互动来实现的。所使用的材料包括樱桃木、胡桃木、石灰石和磨砂玻璃。在室内引入了各种可移动的面（如滑门和旋转门）进一步丰富了在该阁楼中的两个功能区之间对光、空间和肌理的感知。

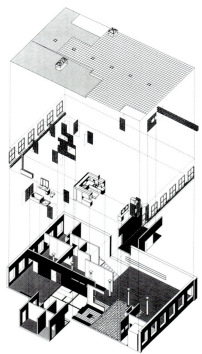

上图及对页图：这个原来的百货商店现今成为了一个注册标志性建筑，其中的古典式立柱是其原有的特征之一，与各个由樱桃木、胡桃木和磨砂玻璃所构成的现代主义式的平面形成鲜明的对比

左图：轴测展开图

ABOVE AND OPPOSITE Classical columns—an original feature of this former department store, now a registered landmark building—offer a counterpoint to inserted modernist planes of cherry, walnut, and frosted glass.

LEFT Exploded axonometric.

上图：该阁楼空间表现出极简主义的特征，胡桃木和磨砂玻璃构成平面和体量。白色乙烯基沙发和椅子在这个当代阁楼中显露出一种突出的雅致

对页图：对材料、细部，以及装修的处理在厨房中最为明显。法国石灰石地板、樱桃木橱柜、以及不锈钢都使这个纯粹功能性的房间变得高雅起来

ABOVE Walnut and frosted glass create plane and volume in this otherwise minimal loft space. The white vinyl couch and chair provide stark elegance within a contemporary architectural envelope.

OPPOSITE The manipulation of materials, details, and finishes is most evident in the kitchen. French limestone floors, cherry cabinetry, and stainless steel elegantly transform an otherwise purely functional room.

Fifth Avenue Loft
SCOTT MARBLE · KAREN FAIRBANKS ARCHITECTS
第五大街阁楼
斯科特·马波及卡伦·费尔班克斯事务所

通过对各种透明、半透明和不透明材料的使用，这个位于第五大街下区的阁楼对紧凑的空间和难得的自然光予以了充分的利用。房主是一名舞蹈设计师，按他的要求，这个面积为1400平方英尺（合130m²）的改造工程包括了两个卧室和两个卫生间，均沿着一系列的从入口到开窗墙体纵向布置的动态平面进行组织和布置。构成这些平面的材料从玻璃棒和各种平板玻璃到滑动式饰面胶合板和玻璃做的隔板，各不相同，从而富有变化。这些平面被用来容纳私密性的和公共性的居住功能。客房位于厨房的旁边，推开滑动隔板后即可作为一个用餐区。而在关上樱桃木隔板时，客房就能够与卫生间联系在一起。各种现代时期的家具给这个原来的工业空间带来温暖感。

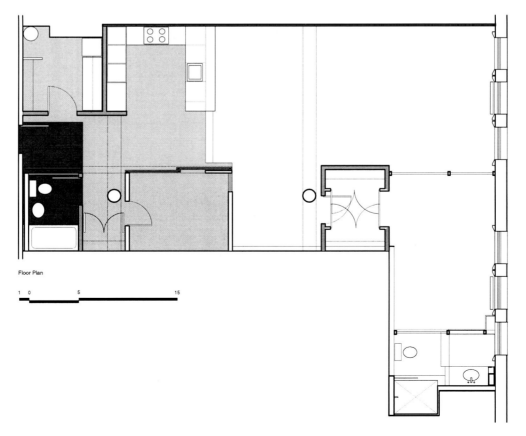

左图：平面图

对页图：三扇窗户是在这个闹市区中相对狭小的阁楼里的两个可以获得自然采光的地方之一

LEFT Floor plan.

OPPOSITE Three panels of windows are one of only two sources of natural light in this relatively small downtown loft.

上图：半透明的隔板给卧室带来一种明亮的感觉

对页图：左侧的滑动隔板可以在招待客人时将客房变成一个正式的用餐区

ABOVE Translucent panels provide a sense of light and brightness in the bedroom.

OPPOSITE Sliding panels to the left allow the guest room to be transformed into a formal dining area for entertaining.

Divney Residence
HUT SACHS STUDIO

戴夫尼住宅
哈特·萨克斯工作室

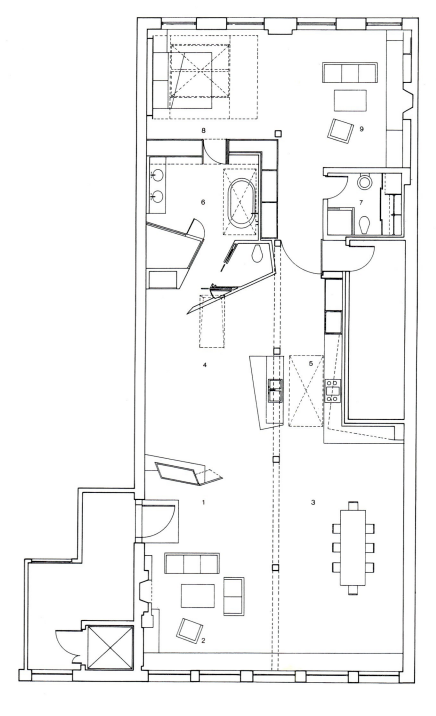

戴夫尼住宅位于市中心，设计师们遵循恢复旧貌和改造创新并重的思路，把它当做具有历史意义的场所而倍加珍惜。业主对于精美手工艺的喜爱在这间2300平方英尺（合213.7m^2）的住宅中精美的细部上获得了共鸣。废弃的砖块被重新用来砌成起居室里的壁炉，而起居室中原有的木梁和木柱子被暴露在外。在整个空间里设置了大量的樱桃木橱柜和金叶饰件，并辅之以定做的面砖和其他一些手工雕刻的木构件。建筑师让这座旧商业建筑的原始性与新建建筑上的精美造型之间形成一种和谐的关系，同时又保留着它们各自的特点。

LEFT Floor plan.

OPPOSITE Cozy living area features a fireplace crafted from salvaged bricks.

左图：平面图

对页图：舒适的起居区中的壁炉用废弃砖块砌筑而成

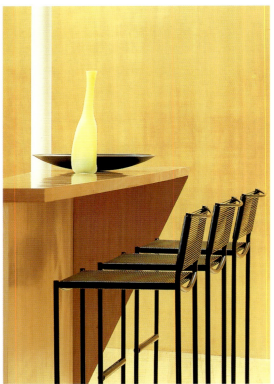

上图：主卧室内的采光天窗和木板面，没有使用装饰性的物品来进行处理

左图：委托人对手工艺品的爱好在阁楼的各个细部的处理上都表现得非常明显

对页图：暴露在外的厚重木梁，新做的樱桃木橱柜，以及枫木地板都沐浴在从头顶上的大型采光天窗处投射而下的阳光之中

ABOVE Master bedroom features skylight and wood planes, limiting the need for decorative objects.

LEFT The client's affection for handcraftsmanship is in evidence throughout the details of the loft.

OPPOSITE Exposed heavy timber beams, new cherry cabinetry, and maple flooring are bathed in daylight from large overhead skylights.

Hudson Loft
ALEXANDER GORLIN ARCHITECT

哈德森阁楼
亚历山大·戈林建筑师事务所

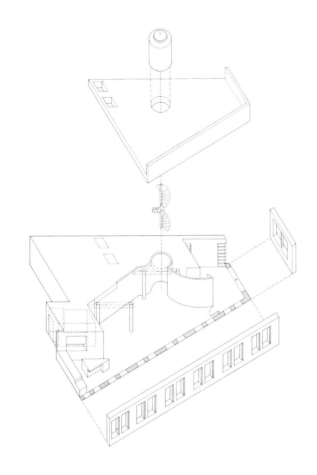

看似昂贵的材料和不规则的平面是这个阁楼的一个主要特色。该项目由一个年轻的家庭所委托设计。室内的台面为不锈钢制作,成本仅为石材的一半。橱柜的表面看上去是用黑檀木制作的,但实际上仅是层板而已。一面由内凹和凸起的曲线所构成的流动墙面,它的复杂的结构和起伏的曲线贯穿了这个平面为三角形的阁楼。主起居区的一端是一个具有后工业气息的厨房,朝西面开窗。中间部分的设计保留了原来建筑的痕迹。在主起居室的另一端,曲面墙体内设置了一个螺旋楼梯,可以一直通到屋顶花园。该阁楼设计的最精妙之处是位于旋梯另一侧的主卫生间。洞穴式浴盆和淋浴器从楼梯侧的凹曲面墙体上塑造出来的,墙面上用深蓝色的瓷砖贴面。

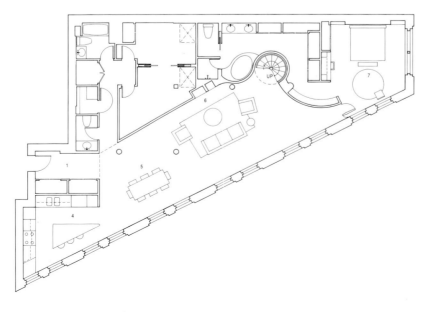

上图:轴测展开图

左图:平面图

对页图:起居区里的流动曲线墙面上设置了书架,并与一个通向屋顶花园的螺旋楼梯相映成趣

ABOVE Exploded axonometric.

LEFT Floor plan.

OPPOSITE Sinuous curving walls in the living area act as bookcases and play against a spiral staircase to the rooftop garden.

上图：主浴室被藏在一个复合曲面中，该曲面也构成了起居室墙面的一部分

右图：具有后工业特征的厨房采用不锈钢橱柜，可以反射从西向的窗户射入的光线

ABOVE Master bath is tucked behind a compound curve that forms part of the living room wall.

RIGHT The postindustrial kitchen features stainless-steel cabinetry that reflects light from the west-facing windows.

Wei Loft
BAK ARCHITECTURE

魏家阁楼
巴克建筑事务所

　　明亮的光线是这个位于纽约市的宁静的阁楼空间最具有特色的地方。这个阁楼既给人丰富多彩的感官享受，同时又显得非常的谦逊节制。它通过采用统一的手法和细微的感觉差别来营造一个日常生活空间。用桦木做的墙板和布艺屏风设置在预先存在的体量之中，用以暗示空间的转换以及在某个围合空间的周边墙面之外的光源。设计师们通力合作，创造出由各个基本的房间所构成的特殊布局结构，并巧妙地将主人的物品进行摆放。这种创造性在书房里体现得最为明显，两个独立式的书架与固定式书架墙巧妙地合为一体。半透明的隔板的表面用帆布铺贴，分别有三个不同的地方采用了这种处理手法。固定式和滑动式屏风被用来营造私密性空间，同时使整个空间的光线显得生动和纯净。

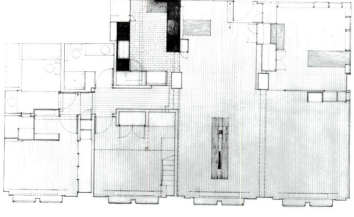

上图：从厨房看室内，图中表现出半透明的隔板和崭新的橱柜

左图：平面图

对页图：入口和书房的内景。通过专门的设计、镶嵌有桦木的墙面、以及半透明的屏风来营造出秩序感和私密性

ABOVE View from kitchen includes trans-lucent panels and new cabinets.

LEFT Floor plan.

OPPOSITE View of the entry and library. Order and privacy have been created by custom-designed, birch-inlaid walls and translucent screens.

上图：主卧室里充满了自然光

对页上图：流动的空间将厨房与起居区联系在一起

对页下图：局部平面图及透视图

ABOVE The master bedroom is filled with natural light.

OPPOSITE ABOVE Flowing space links kitchen to living area.

OPPOSITE BELOW Partial plan and perspective view.

City Loft
VICENTE WOLF ASSOCIATES, INC.

城市阁楼
维森特·沃尔夫联合公司

该阁楼中的家具和艺术品来自法国、瑞典、中国、泰国等国家，它们非常个性化地表明了业主兼设计师的独特品位以及对旅游的爱好。中性的色调以及简易、空敞的室内空间成为本设计的主要特色。室内部分通过采用各种尺度的搭配以及一些独立式的物品来获得平衡的效果。黑色和白色家具以及黑白照片所形成的一种折中式的混合与漆成白色的混凝土楼板交织在一起。中性色调的石材、亚麻布和皮革形成了阁楼空间中各式各样的质感。工艺品不是挂在墙壁上，而是摆放在重要位置，从而能够抓住并吸引人们的注意力。周围的城市景色成为内部空间的点缀，业主兼设计师让窗户保留原样，没有进行任何的处理。

上图：桌子和凳子是室内独立式家具的一部分，丰富了该阁楼在尺度上的多样性特征

对页图：起居区里的中性家具、刷白的墙面，以及刷白了的混凝土楼板

ABOVE Desk and stool are some of the freestanding pieces that add to a variety of scales in the loft.

OPPOSITE Living area features neutral furniture, painted walls, and a concrete floor painted white.

OPPOSITE Furniture and objects from around the globe reflect the owner-designer's love of travel.

RIGHT AND BELOW White is dominant in this bedroom, which is accented with black and white photographs.

上图：来自世界各地的家具和物品反映出业主兼设计师对旅游的爱好

右图及下图：在该卧室中，白色成为主题，黑白照片进一步强化了这一效果

Chelsea Loft
KAR HO ARCHITECT

切尔西阁楼
卡尔·霍建筑师事务所

这个面积为1200平方英尺（合111.5m²）的切尔西阁楼是业主兼设计师的收藏品的展览橱窗，这些收藏品包括色彩明亮的家具和1960年代以后的一些装饰品。该阁楼空间首先被拆到只剩下木地板和周边的墙体，然后采用白色涂料饰面，吊顶有9英尺（合2.7m）长。建筑师面临的最大挑战是要从建筑北侧的一个大窗户处将自然光引入到空间的内部。为了获得最大限度的光亮感和营造空间的深度感，建筑师专门设计了用磨砂玻璃做的屏风和门扇。由于主人是独自居住，所以没有必要采用墙壁来营造一种私密感。卫生间很长，阳光无法照射到里面去，所以再一次采用了磨砂玻璃来满足对自然光线的向往。

上图：磨砂玻璃的入口门将间接光线引入厨房

对页图：通过少量的分隔墙和明亮的白色涂料，在这个小小的阁楼里实现了一种宽敞的感觉

ABOVE Frosted-glass entry door brings indirect light into the kitchen.

OPPOSITE A feeling of spaciousness is achieved in this small loft through limited dividing walls and fresh white paint.

ABOVE As the owner lives alone, the architect was free to create an open plan.

LEFT Frosted-glass mirrors in the bathroom create the sensation of windows in a long, narrow room otherwise lacking in natural light.

OPPOSITE Brightly colored objects are set off against the neutral tones and large volume of the loft.

上图：由于该业主独自居住，因此，建筑师可以自由地营造出一个开放式的平面

左图：浴室里的磨砂玻璃镜子在一个狭长的、原本缺乏自然采光的房间里形成了一种窗户的感觉

对页图：色彩明亮的物品被该阁楼里的中性色调和巨大的体量衬托得非常醒目

右图：起居室刷成白色，为业主收藏的 1960 年代的家具提供了一个宁静的背景

RIGHT Sitting room painted white provides a serene backdrop for the owner's collection of 1960s furnishings.

Rosenberg Residence and Studio
BELMONT FREEMAN ARCHITECTS

罗森伯格住宅与工作室
贝尔蒙特·弗里曼建筑师事务所

要想在住宅里分隔出工作空间通常是非常困难的事情。在本项目中,房主拥有一个令人倾羡的条件:两个房间垂直叠加在一起,其中一个位于另外一个房间的顶上。在改造这两个面积为 1500 平方英尺(合 139.4m²)的楼层时,建筑师们既强调了相互间的垂直划分,也发展了一套材料上的语言,将两个楼层统一起来。

虽然外面的城市有着迷人的景色,但阁楼的内部空间却显得平静而安祥。建筑师们强调了起居楼层和工作楼层之间的相互联系,但并没有采用普通的楼梯,而是采用了轮船的舷梯,其坡度达到了 70°,象征着那种富于挑战性的攀登。灯光的运用进一步形成了两个平面之间的差异:工作空间采用了具有商业性特点的活动轨道灯和一组由威廉·勒斯卡兹(William Lescaze)设计的吊灯;而在住宅的主要部分采用了落地灯和台灯的搭配,形成了独特的效果;而在狭长的白色厨房墙壁部分采用了非直射式的荧光灯来照亮厨房。

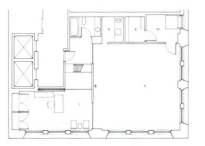

上图: 从曼哈顿大街看该建筑,显示出在黄昏十分有两层被照亮的情景

左图: 平面图,楼上的起居区与在楼下的工作室

对页图: 船上用的楼梯将两层连接起来,同时在工作和起居区之间形成了一个突出的转变。各个用冷轧钢制作成的门用车蜡来进行最后的装修处理

ABOVE A view of the building from the Manhattan street shows the two levels illuminated at dusk.

LEFT Floor plans, living area above and studio below.

OPPOSITE A ship's ladder connects the two levels while providing a significant transition between work and living areas. Cold-rolled steel doors have been finished with automotive wax.

LEFT The upper level includes the living room, kitchen and two bedrooms. The owner's collection of mid-century modernist furniture is featured throughout.

BELOW On the lower studio level, the original concrete floors have been refinished, as have the radiators, which were removed, sandblasted, and sprayed with molten zinc.

左图：上一楼层中设置了起居室、厨房和两个卧室。在各个地方都布置了该项目的业主所收藏的上世纪的现代主义家具

下图：在下面的工作室层里，原有的混凝土楼板被重新处理，室内的散热器被卸下来，重新进行了喷砂和镀锌处理

上图：采用悬浮着的带抽屉枫木台面来强化不锈钢厨房的效果。原来的混凝土楼面进行了重新装修处理

ABOVE The stainless-steel kitchen is accented by a floating maple countertop box with drawers. The original concrete floor has been refinished.

右图：盥洗室成为封闭的淋浴间的入口处。该房间用2英寸（合5cm）厚的雪松木做吊顶以吸收潮气。不锈钢的水槽从钢制支架上悬挑出来

RIGHT The washroom becomes the lobby area for the concealed shower/wet room. This room features a 2-in-thick (5-cm-thick) cedar ceiling to contain moisture. The stainless-steel sink is cantilevered on steel braces.

Live/Work Dualities Loft
DEAN/WOLF ARCHITECTS

生活/工作两用阁楼
迪安/沃尔夫建筑师事务所

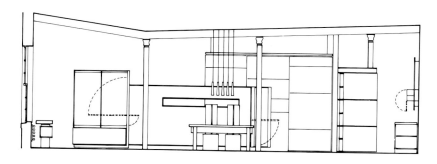

充满阳光的顶层空间是这个小小的特莱贝卡阁楼原来最吸引人的地方，而它被改造成了一个供人生活和工作的场所。对于这个 875 平方英尺（合 81.3m²）的敞开空间，设计的方针是创造出一个由可移动构件组合而成的"可操作的拼贴画式"的系统。在工作间、卧室和浴室三个空间之间，视觉上的联系在实际上和人的感受上扩大了的空间的尺度。

这个阁楼空间的外围结构未加变动，仅对它的铸铁柱、木地板和石膏粉刷墙做了翻修和粉刷。新加建的部分设置在这样一个中性的背景中。每件新加建的要素都采用混凝土、钢材和重木作为结构构件。与这些粗糙的材料并列放置在一起的是精致的纤维材料——帆布、棉、毛以及铝丝网，从而增强了家庭生活的气氛。

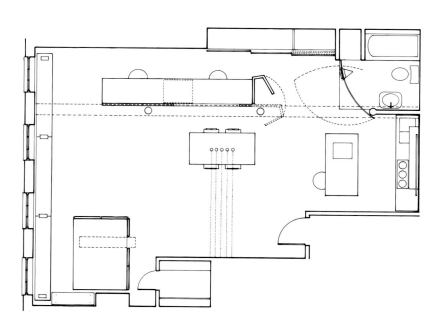

上图：剖面和平面

ABOVE Section and floor plan.

对页图：工作室墙面和空缝隙处的细部。并列放置的材料增强了对光线的反射。桦木胶合板和铝丝网屏风将卧室从厨房/用餐区独立分隔开来

OPPOSITE Detail view of workroom wall and aperture. Juxtaposed materials enhance light reflection. A birch-plywood and woven-aluminum screen isolates the sleeping quarters from the kitchen/dining area.

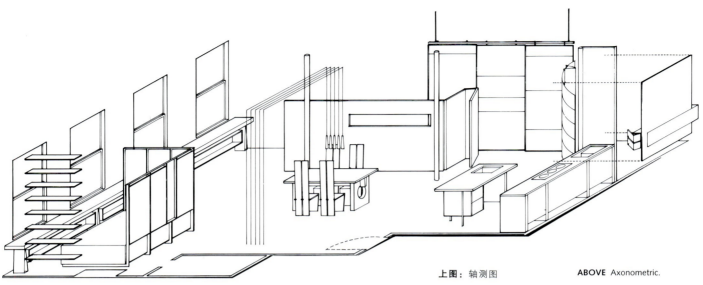

上图：轴测图

左下图：钢材与帆布屏风

右下图：前景为用帆布与钢材做成的浴室门，可以看到生活区

ABOVE Axonometric.

BELOW LEFT View of steel and canvas screen.

BELOW RIGHT Canvas and steel bathroom door in foreground with view to living area.

ABOVE LEFT AND RIGHT
Cherry and steel chairs in the dining room, with the workroom wall and translucent screen beyond.

左上图和右上图：餐厅里的樱桃木和钢制椅子，后面是工作室的墙壁和半透明的屏风

Artist's Studio
GALIA SOLOMONOFF PROJECTS
艺术家工作室
加里亚·索罗莫诺夫设计室

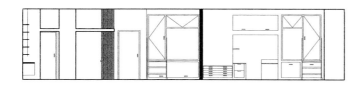

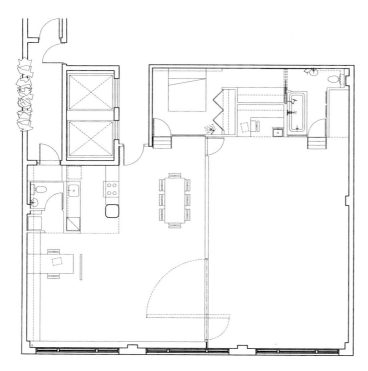

该阁楼位于一座修建于的1910年的毛皮厂房内，它当初仅有一部已经坏掉了的货运电梯、一个已经不能用的浴室、还有就是厨房里的一个摇摇欲坠的旧餐具室。这幢建筑为钢筋混凝土结构，有着巨大的梁和柱。虽然预算非常紧张，但户主兼设计师在不到10个星期的时间里将这个2000平方英尺（合185.5m²）的阁楼改造成了一个住宅。设计师的丈夫是一位艺术家，他有一个明确的要求：他希望在躺在一个蓝色的浴缸里的时候能够看得到工作室里他自己的绘画作品。按照他们自己的愿望，这对夫妇从工作室和起居区之间将这个48英尺×44英尺（合14.6m×13.4m）的空间分成两半。这个阁楼尽量满足工作和家庭生活的需要，在靠北向墙壁处设置了一个抬高了的平台，安排了主人的床、浴室和其他的一些私密要素，同时还提供了一个宽敞的工作区域。在住宅的室内处理上很少采用的成品工业材料，包括玻璃、铝材和其他标准的商品，成了这个设计精美的阁楼的中心。

上图：平面图

对页图：图书房里用钢筋将钢丝网做的隔板吊在顶棚上。在原有的窗户旁设置了一个躺椅，散热器用钢丝网罩住

ABOVE Floor plan.

OPPOSITE A dividing wire-mesh panel in the library is hung from the ceiling slab using steel rods. Original windows feature a bench flanking the radiators, screened under wire mesh.

ABOVE Beyond the dining area a stepladder leads to a raised sleeping platform, which can be closed off for privacy, with bathroom and walk-in closet.

OPPOSITE Domestic functions of kitchen-laundry-pantry are efficiently packed under a light aluminum canopy in an area that was formerly the elevator shaft. The refrigerator and stove were found in a second-hand restaurant supply shop in the Bowery.

上图：在用餐区的后面有一部活动梯通向一个抬高了的睡觉平台，它可以被封闭起来以实现对私密性的要求。它带有浴室和步入式橱柜

对页图：厨房、洗衣房、餐具室等家庭生活的功能被紧凑地压缩在一片轻质铝天棚下，原来这里曾是一个电梯井。冰箱和炉子都是从位于鲍威利区的一家二手厨房设备商店里买来的

Tribeca Home and Studio
MONEO BROCK ARCHITECTS
特里贝卡住宅与工作室
莫内欧·布洛克建筑师事务所

这个在闹区的阁楼位于一幢1898年建成的10层楼仓库的顶层上。为了保持一种宽敞的感觉，设计师在一些重要的部位将屋顶切开，并安装了天窗或宽大的高侧窗。在主要起居区部分的玻璃让光线能够照到空间的深处，还可以看到在该建筑屋顶上的水箱。折叠式船用铝质楼梯用来作为通向屋顶花园的通道。该楼梯还设有一个用玻璃砖做成的平台，并将楼梯做得很轻，很容易将梯子提上去，从而让出更多的工作空间。在阁楼的周围，一些仓库原有的柱和梁保持不变，以加强墩实感。定制的带毛玻璃的橱柜将工作区与生活区分隔开来。

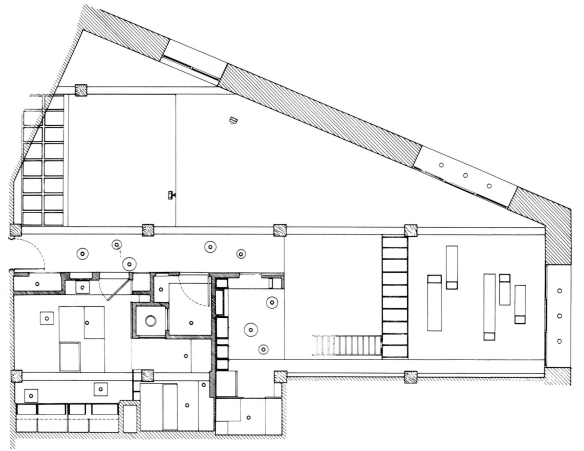

左图：平面图

对页图：从工作区透过阁楼看天窗下的船用楼梯。压制板墙为厨房提供了私密性

对页右图：在用餐区上方，一面大型侧高窗朝向屋顶和水塔

LEFT Floor plan.

OPPOSITE View from the work area through the loft to skylit ship's ladder. Pressboard wall provides privacy for kitchen.

OPPOSITE FAR RIGHT Above the dining area, a majestic clerestory wall opens onto the rooftop and water tower.

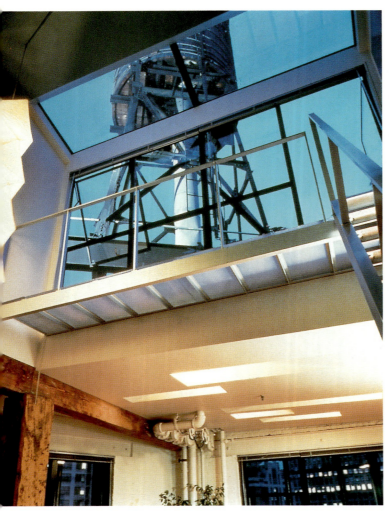

ABOVE Frosted-glass landing bridge allows a close-up look out to the cityscape and skies above.

ABOVE RIGHT The romantic rooftop garden sits under the water tower.

上图：磨砂玻璃做成的天桥平台构成了一个特写镜头，将城市景象和天空尽收眼中

右上图：富于浪漫气息的屋顶花园设置在水塔的下方

左上图：厨房范围很小，但功能完备，设置了经过精心设计的橱柜

上图：工作室有着高敞的吊顶、巨大的天窗，以及充足的自然采光

ABOVE LEFT Kitchen area is small but functional, with carefully designed cabinetry.

ABOVE Working studio features high ceilings, large clerestories, and plenty of natural light.

GOING GLOBAL
走 向 世 界

在一个资源日益减少的世界里，废弃建筑物的重新利用已经成为世界各国优先考虑的问题。对现有建筑的不满使许多人希望在既有的习俗范围内建立一种新的秩序。对于这些愿望，阁楼可以说是一座绝佳的宝库供人们去发掘。在伦敦，新潮人物们正在对原来的商用造船厂的一些建筑物进行改建。在东京这样的空间永远紧张的地方，阁楼开发项目正在一些无任何特征的普通仓库建筑中发展起来。随着新兴商业城市不断要求增加更为新颖的设施，城市的政府部门在推动将老图书馆或一些建筑改造成为居住阁楼项目的时候也看到了其中的好处。

在某些人来说，阁楼只不过是混凝土外壳加上一些未加饰面的楼板和墙面。但是，一个差不多是冷冰冰的敞开空间却为那些有创造性眼光的人提供了表现自己的机会。虽然那些守旧的人断言，阁楼即使配上黑色皮革、不锈钢、以及几幅黑白照片，它仍只不过是阁楼而已，但阁楼居住方式却真正要抛弃所有那些传统的观念。在进行室内规划和家具安排时，要知道从哪里着手并不是一件容易的事情，但不管怎样，最重要的一点就是舒适性。

阁楼设计的主要挑战之一是将建筑物的过去和现在结合在一起。为了达到这种统一，世界各地的阁楼往往像壁毯那样，需要将那些既矛盾又有联系的东西编织在一起。在设计中经常会受到现代主义追求理性设计和秩序的影响。但在追求对粗犷的工业建筑面貌的超越中，最终的阁楼往往会让人们的愿望落空，在传统的片段与纯粹的、开敞的空间之间形成了一种妥协，对功能要求和大胆的想像会同时出现在同一个空间之中，形成一种个人的乌托邦式的感觉。在阁楼设计中，设计师和建筑师将客户推向了在材料应用和表现手法上的极致，同时，它也会受到人们在对阁楼的稳定性和安全性上的要求的影响。

在整个世界范围内，有关阁楼居住方式的设计中最值得一提的是，这些设计往往会拒绝当地的文化形式。阁楼居住方式含蓄地认同了个人和我们要创造具有与社会生活不同的意义的愿望。因此，一个阁楼或许就是地球村和世界居民们真正的居所。在与潮流和传统的诱惑所作的抗争中，后面将要介绍的那些阁楼在全球不同的文化环境背景下提供了一种属于当代的住居方式。

LOS ANGELES

Carlson-Reges Residence
RoTo ARCHITECTS, INC.

洛杉矶

卡尔森—里杰斯住宅
洛托建筑师事务所

洛杉矶市所具有的个人崇拜特征常常会在一些桀骜不驯的建筑项目上表现出来。以洛杉矶为基地的洛托建筑师事务所的合伙人迈克尔·洛汤第和克拉克·史蒂文斯在靠近市区的一个原有工业带里将该城市的第一座发电厂的机械车间进行了改造。该建筑有着引人注目的建筑形象,成为一个独具特色阁楼。

在保留结构原有的粗犷、开敞的特色的同时,该建筑被改造成一个相当舒适的地方。建筑师采用了一个在新的和旧的部分、在内部和外部、在上面和下面之间来回跳跃的几何形式,从而模糊了在空间划分上的边界。混凝土和钢材与红木平台和一个茂密的热带雨林花园并存。窗户和门框被拆除掉,以便将新的钢构件放入建筑中,在原有的围护墙上形成很大的开口,从而室内和室外空间能够相互畅通。

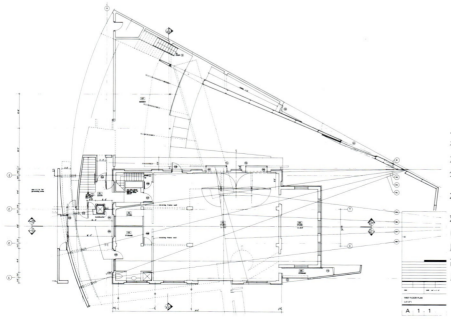

上图:建筑外部重新构图,在原有混凝土与钢骨架顶上加一个钢壳,其下为卧室部分。用角钢板支撑屋面梁架

ABOVE The exterior was reconfigured and a steel shell was added atop the existing concrete and steel pavilion, which contains the bedroom annex. Angular steel panels support rooftop beams.

左图:平面图

LEFT Floor plan.

对页图:空间高达36英尺(合11m)的起居室位于地面层。上面楼层的空间也是很高,但布置得更具私密感,用作主卧室,而夹层是客房

OPPOSITE Vertiginous 36-foot-tall (11-meter-tall) living room occupies the ground floor. The upper floor is the still soaring but more intimate setting for the master bedroom and mezzanine-level guest room.

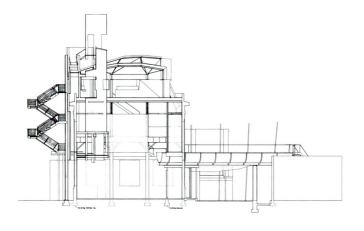
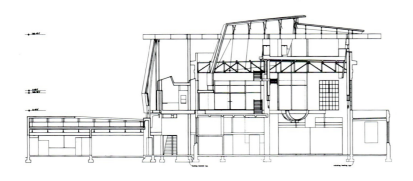
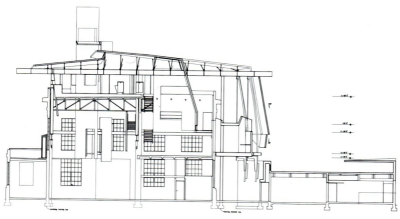
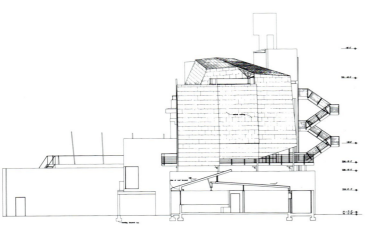

ABOVE On the ground floor is a semiprivate gallery for the owner's art collection.

LEFT Sections.

OPPOSITE Stairway leads from living area to the bedroom annex.

OPPOSITE FAR RIGHT The architects did not simply create a geometric collage of found parts. They extrapolated lines from the surrounding landscape, landmarks such as Dodger Stadium, mountain peaks, and the downtown skyline influenced the interior design.

上图：地面层为半私密的陈列室，摆放着业主的艺术收藏品

左图：剖面图

对页图：楼梯从起居区延伸到卧室部分

对页右图：建筑师们不仅仅是把现有东西按几何方式拼接起来。他们亦将周围的景观、像道奇体育场这样的地标性建筑，山峰，以及闹市区的天际线等都引入到阁楼之中，对室内设计的效果产生了很大的影响

CARLSON-REGES RESIDENCE 93

CHICAGO

Concrete Loft
FRANKEL + COLEMAN

芝加哥

混凝土阁楼
弗兰克尔 + 科尔曼建筑师事务所

这座全混凝土的阁楼位于芝加哥具有历史意义的普林特斯罗区（Printer's Row District）中央的一座标志性建筑里。其主人夫妇都是建筑师，他们对尝试在开放场所中的起居生活有着浓厚的兴趣。这个阁楼的主要特点是采用了混凝土楼板和顶棚，以及一些旋转门，只有两个卫生间是全封闭的。由于主人夫妇俩都有很多时候是在家里工作，所以这个阁楼成了在空间设计上的试验，需要对它进行改动和调整，以满足家庭生活和工作的需要，并且要便于主人们随时可以看到对方。在室内设计上的斯巴达式的简洁布置使这个获奖的 3500 平方英尺（合 325.2m²）的阁楼有一种宽敞的感觉。主人还是 20 世纪现代派家具和艺术品的狂热收藏者，他们定期地将这些收藏品从仓库轮流搬进和搬出这个阁楼。建筑师们只建起了少量的墙，并对窗户进行了处理，从而将繁忙的城市景色挡在视线之外，实现了极简主义式的效果。

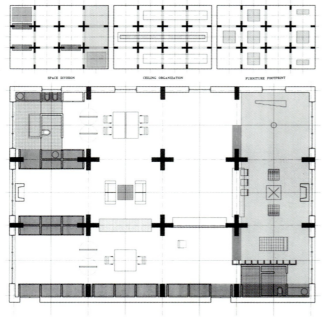

上图：该阁楼位于在芝加哥市中心的一座有历史意义的建筑里

左图：平面图

对页图：阁楼组织在 12 个方形开间中。这个 3500 平方英尺（合 325.2m²）的主要特色是家庭用房和厨房区，其中包含一个内置图书馆和不锈钢工作台

ABOVE Loft occupies space in a historic building in the center of Chicago.

LEFT Floor plan.

OPPOSITE The loft has been organized around twelve square bays. A central feature of this 3,500-square-foot (325.2-square-meter) space is the family room and kitchen area, which includes a built-in library and stainless-steel countertops.

左图：用带质感的涂料来涂刷原来的混凝土柱子，让其与混凝土楼面和新加建的电镀管道之间形成良好的配合效果。室内的背景是勒·柯布西耶式的沙发和克诺尔式的椅子

对页图：沿室内中央轴线看，远处是独立式煤气壁炉

LEFT Original concrete columns have been painted with textured paint to complement concrete floors and new galvanized duct work. Le Corbusier sofas and Knoll chairs (background).

OPPOSITE Long view down central axis to freestanding gas fireplace.

ABOVE The rugged kitchen features stainless-steel cabinets and commercial appliances.

OPPOSITE The working area looks through pivoting doors to the baby's quarters, which feature a custom-designed, stainless-steel crib and Alexander Calder mobile.

上图：粗糙的厨房凸显出不锈钢橱柜和商业化的用具

对页图：从工作区穿过旋转门可以一直看到婴儿活动区，在其中安置了一张专门设计的不锈钢婴儿床，以及亚历山大·考德尔式的活动玩具

ANTWERP

Coolen House
KRIS MYS ARCHITECT

安特卫普

库伦住宅
克里斯·迈斯建筑师事务所

　　欧洲的阁楼设计手法在有5层的库伦住宅的改建工程中走出了一条新的道路。该住宅是一栋1905年建成的新艺术风格建筑，位于比利时安特卫普市。该工程的焦点问题是如何将自然光引入到这个原来阴森黑暗的住宅里去。第一项改造工作就是将非承重墙拆除，让5层楼相互敞开，为完成这项工作，建筑师还改变了层高。从顶层可以一直看到建筑的底层。5层高的中庭同时也作为陈列空间来展示珍贵的收藏品。在这栋装饰华丽的建筑物的后面的扩建部分被用一个矩形盒子来替代，盒子还有一个短翼，转了一个角度挑出。该阁楼容纳了主人的办公室，它仿佛一直延伸流动到花园中去了。

左图：该室内的中心点是带有玻璃格架的锈红色钢结构

对页图：在该建筑中修建了一个专门设计的钢楼梯。层高有所改变，并且墙体也被拆除了

LEFT The focal point of the interior is a rust-colored steel construction with glass casing.

OPPOSITE A custom-designed steel staircase was built into the building. Floor heights were altered and walls removed.

上图：这个阁楼式的住宅成了一个橱窗，专门展示露天市场上买来的有价值的收藏品

对页图：钢和木制的桁架，以及铝板覆面的保温墙体，将这个黑暗的阁楼改变成了一个光线充足的卧室

ABOVE This loftlike residence is a showcase for a valuable collection of fairground figures.

OPPOSITE Steel and wood trusses and aluminum-backed insulation converted the dark attic into a light-filled bedroom.

LONDON

Oliver's Wharf
MCDOWELL + BENEDETTI

伦敦

奥立弗码头
麦克道尔 + 贝内迪提事务所

在伦敦塔桥的下游仅几百米处，从前的茶叶仓库被改建成了豪华的奥立弗码头阁楼。这个项目由一个年轻的设计小组按时新的方式对它作了处理。这个富有特色的屋顶阁楼位于首都的这个讲究时髦的道克兰社区里，设计者们希望在同一栋楼里既要体面豪华，又要表现出极简主义的特征。客户提出的任务要求非常具体：最大限度地扩大朝河的视线，以及在尽可能保持该空间的开放性的同时创造出一块进行烹调、娱乐和艺术活动的场地。建筑物的大部分都损坏了，原有的木构件和砖头都暴露了出来。为了营造一种大都市的环境气氛，设计人员将现代材料和原有的元素混合在一起。加上了铸铁柱来支撑橡木桁架，而墙面也做了喷砂处理。在整个阁楼里都安装上了彩釉面钢制屏风、橱柜门和压铸玻璃板。顶部处理成一个屋顶平台，可以360°地欣赏泰晤士河的景色。

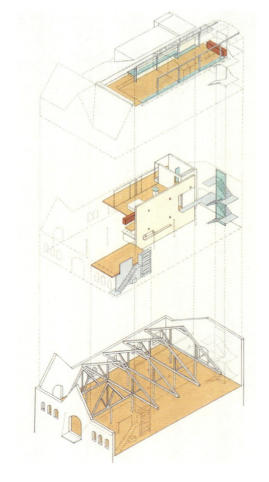

上图：奥立弗码头建于1870年，是一个受到保护的伦敦标志建筑。在一个世纪之后，它成了这个城市首批阁楼改造工程之一

左图：展开的轴测图显示新的要素是如何被插入到原有结构中去的

对页图：壮观的屋顶小亭可以将泰晤士河沿岸的景色尽收眼底

ABOVE A protected London landmark, Oliver's Wharf was built in 1870. A century later it has become one of the city's first loft-block conversions.

LEFT Exploded axonometric diagram shows how new elements were inserted into the existing structure.

OPPOSITE Spectacular rooftop pavilion offers windswept views along the Thames.

ABOVE A cast-glass stair screen illuminated from above by natural light and from below by uplights.

LEFT Enameled steel kitchen cabinets were designed to maximize light.

上图：室内设置了一面玻璃楼梯屏风，上面用自然光线来照明，其下部用投光灯来照明

左图：抛光的钢制厨房橱柜的设计目的是为了最大限度地获得光亮的效果

ABOVE The bedroom, where there appears to be nothing between the bed and the Tower Bridge but a sheet of glass.

上图：卧室内景，在这里看上去在床和塔桥之间除了一片玻璃外似乎没有任何东西

RIGHT Two-story fireplace screen in the main living area contains not only a gas fire and bookshelves but also conceals a steel stairway.

右图：主起居区里两层高的壁炉屏风上不仅设置了一个煤气壁炉和一些书架，还遮挡住了一个钢制的楼梯间

AVOCA, PENNSYLVANIA

O'Malley Residence
CARPENTER/GRODZINS ARCHITECTS

宾夕法尼亚州,阿沃卡

奥马利住宅
卡彭特/格罗津斯建筑师事务所

这个优雅的奥马利阁楼面积为1600平方英尺（合148.6m²），位于宾夕法尼亚州斯克兰顿市郊区的一个小镇上。该建筑原为一座仓库，改建时将其拆除得仅剩外围的墙体、柱子和梁，最终留出一个完整的容器来容纳主人所提出的一个很小的任务要求：完全开敞的空间。该阁楼是为一位单身者设计的，因此，对私密性的要求不高，让建筑师可以保持空间的敞开性和透明性。建筑师考虑的首要问题是最大限度地实现自然采光。柱子、墙和镶嵌在橡木地板里的大理石边带有助于形成一种随意的秩序，并形成在居住区、用餐区和睡眠区之间的划分。厨房和浴室顺着朝向西南的带窗墙壁设置，这些窗户提供了充足的自然光。用毛玻璃和透明玻璃做成的形体有助于限定空间和流线模式。

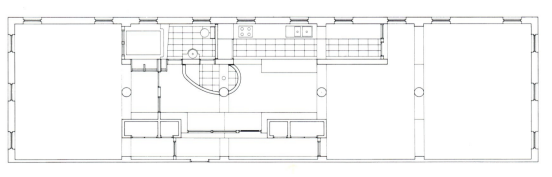

上图：在入口处两侧设置了一对储藏间，正面采用桉木和玻璃，形成室内的墙体并且提供了壁柜空间

左图：平面图

对页图：在整个阁楼中到处设有蚀刻玻璃做成的储物柜

ABOVE Flanking the entry are a pair of ash- and glass-fronted storage units that suggest walls and provide closet space.

LEFT Floor plan.

OPPOSITE Etched-glass storage cabinets are featured throughout the loft.

ABOVE The main living space is dominated by columns and marble inlays in the oak flooring that break up the space. Behind the dining table, a translucent glass panel acts as a dividing wall from the entry.

OPPOSITE The bedroom is part of a seamless continuum with the main living space. The etched-glass front of a storage unit, which bridges the bedroom and entry, provides only hints of the objects within.

上图：主起居空间里的柱子和橡木地板中的大理石嵌条将空间进行了划分。在餐桌的后面，一块半透明的玻璃门扇成为在入口处的一面分隔墙

对页图：卧室成为主起居空间延续的一部分。储藏单元的刻花玻璃立面将卧室与入口联系在了一起，暗示着在卧室里面的处理

LONDON

New River Head
MCDOWELL + BENEDETTI

伦敦

新河源住宅
麦克道尔 + 贝内迪提事务所

这个有两个卧室的阁楼曾经是伦敦市水资源委员会的会议室，建于1921年，至今还保留有原来的华丽装饰。该房间非常宽敞，有两层那么高，同旁边有巨大的科林斯柱式的门厅过道一起，都被精心修复过，而且大部分地方都刷成了白色。会议室被改成了一个巨大的起居室，富丽堂皇的窗户朝向大街。在房间两端的原有阳台被改造成藏书室，有一个狭窄的楼梯可以通往这里，楼梯本身隐藏在新建的独立式隔墙的后面，墙上布置了大幅的艺术作品。设计小组采用了一些当代要素来满足功能的需要，并且与原有的一些精美的细部处理形成对比。阁楼中设置了一个小楼层，安排了两个卧室和一个大卫生间。一个通往卧室的新楼梯用双层钢材制成，在平台处还设置了一个大鱼缸，将它作为栏杆的一部分。

上图与右图： 主起居区内景，它曾经是伦敦市水资源委员会总部办公楼中庄严的会议室

ABOVE AND RIGHT Views of the main living area that was once the majestic boardroom of the London Metropolitan Water Board's head office.

上图：狭窄的楼梯间通向图书馆的阳台，用新加建的独立式屏风墙遮挡起来，在墙面上安放了大型的艺术品

对页上图：在二层上的卧室区非常宁静而简洁

对页下图：设计小组采用了现代艺术品将色彩和戏剧效果引入这个在很多地方显得严肃的阁楼空间

ABOVE Narrow staircases leading to library balconies are concealed behind new, freestanding screen walls that incorporate large-scale artwork.

OPPOSITE ABOVE Bedroom area on the second level is serene and simple.

OPPOSITE BELOW The design team has relied on modern artwork to bring color and comedy into this largely serious loft space.

JERSEY CITY

Pavonia Loft
ANDERS ASSOCIATES

泽西城

帕沃尼亚阁楼
安德斯联合事务所

在帕沃尼亚阁楼设计上的动机是要体现一种在现代社会里追求一种新的生活方式。这个5000平方英尺（合465.5m²）的空间以前是位于新泽西州滨水区面向曼哈顿的一座仓库，被两家人购买了下来。他们在建造住宅的目标上是打算在靠近城区的地方营造出一个在经济上承受得起的住所。由于是两家人占用这个阁楼，在设计上必须采用花费较小的创新方案。主起居空间由两家人分享，它被分隔成两部分，同时也形成了宽敞的公共开放空间。每个家庭都有单独的私密空间，包括一个厨房，一个卫生间和两间卧室，布置在一个共用的洗衣间周围。在特征上，该阁楼体现出在新旧之间的冲突。阁楼里采用的材料并不算贵，包括用玻璃纤维建造的竖塔和常用在地道里的格栅楼板，保持了阁楼本身所具有的严肃的工业建筑原貌和设施。

上图：共用起居区内景。竖塔从内部照明，犹如一个巨大的日本式灯笼

左图：室内透视图

对页图：主起居空间。光线从上面采光天窗通过3层楼高的竖塔透入；镂空格栅楼面通常可以在地下通道里看到

ABOVE View of shared living area at night. Tower is illuminated from within like an enormous Japanese lantern.

LEFT Interior perspective.

OPPOSITE Main living space. Light washes through a three-story tower from a skylight above and open-grate floors commonly found in subways.

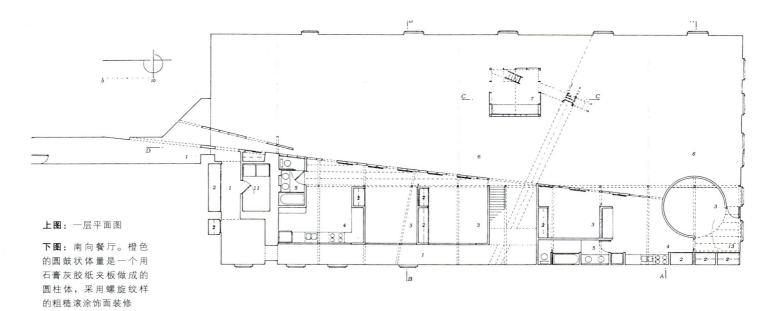

上图：一层平面图

下图：南向餐厅。橙色的圆鼓状体量是一个用石膏灰胶纸夹板做成的圆柱体，采用螺旋纹样的粗糙滚涂饰面装修

ABOVE First floor plan.

BELOW South dining room. Orange drum is a sheetrock cylinder with spiraling rag-roll finish.

右图：南向卧室。天桥边上安放了一些树枝来作为栏杆。爬梯成为从起居室通向天桥的通道

RIGHT South bedroom. Bridge incorporates branches of trees as ballusters. Ladder allows access to bridge from living room.

ABOVE Bridge intersection from below.

OPPOSITE Cantilevered study at night. Red desk faces tilted screen of wood and Plexiglas. Steel duct runs length of loft, approximately 100 feet (30.5 meters).

上图：从一层看天桥交接处

对页图：悬挑出来的书房的夜景。红色的桌子朝向倾斜的用木和有机玻璃制成的屏风。钢制的管道贯穿整个阁楼，大约有100英尺（合30.5m）长

LONDON

Neal's Yard
RICK MATHER ARCHITECTS

伦敦

尼尔码头住宅
里克·马瑟建筑师事务所

那些正在寻找时髦住所的伦敦房地产机构早就对科文特工业园内的仓库和轻工业建筑垂涎三尺了。这一区域环境优美，风景如画，区内有许多狭长的砖房，里面光线不足，空间狭小。在远离喧嚣繁忙的尼尔码头的地方，里克·马瑟建筑师事务所在一座从前的仓库建筑的外围结构上雕琢出了这个壮观的屋顶阁楼。阳光从一个中心筒体上泄下来，整个空间通风良好，十分开敞。客户希望他在淋浴的时候能够看得到天空。设计者安装了顶棚采光的楼板，包括一个当人轻轻触碰开关时会变得不透明的特殊电子玻璃罩，这些处理使人能够观望天空，同时又可以在需要时保持私密性。

上图：顶层的书房隐藏在隔墙后面，但仍然是开敞空间的一部分

ABOVE Top-floor study is hidden behind a partition wall, yet is still part of the open plan.

对页图：原有的楔形仓库被建筑师进行了巧妙的处理。卧室里的采光天窗可以在转瞬之间变得不透明

OPPOSITE The original wedge-shaped warehouse has been cleverly manipulated by the architects. Bedroom skylight can instantly be made opaque.

右图：这个位于屋顶上的阁楼是现代风格的一个巨大的胜利，其中有着充足的自然采光和一批现代家具收藏品

RIGHT A triumph of modern style, this rooftop loft is filled with light and a collection of modern furniture.

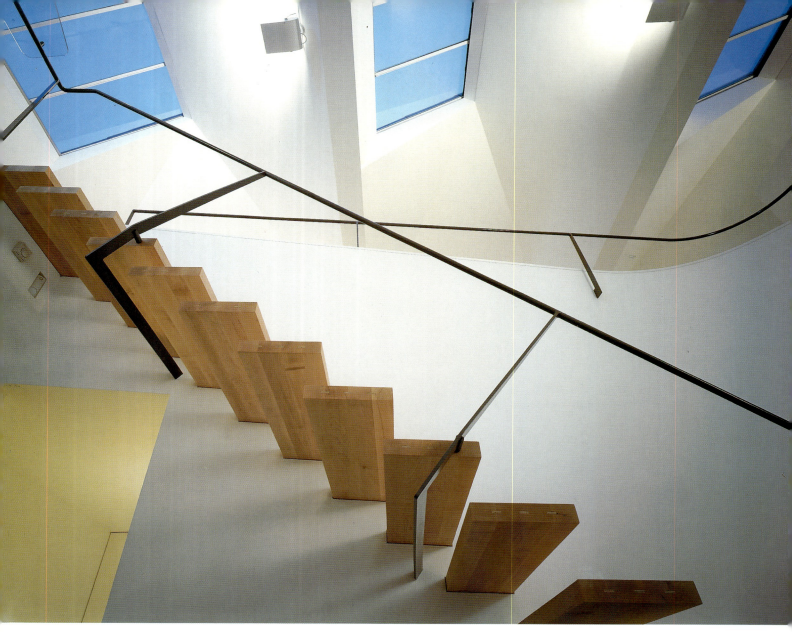

上图：中心核楼梯井将光线从折尺形的采光天窗处倾泻而下，照亮了下面的楼层

对页图：屋顶平台形成了在以烟囱顶和砖砌建筑为主的城市景观中的一点绿化

ABOVE The central-core stairwell allows light from the zigzag line of skylights to pour down into the lower floor.

OPPOSITE Rooftop terrace provides a bit of greenery in this urban landscape dominated by chimney tops and brick buildings.

LONDON

Chiswick Green Studios
PETER WADLEY ARCHITECTS

伦敦

奇斯维克·格林工作室
彼得·华德利建筑师事务所

该阁楼是在西伦敦的再开发工业建筑中的阁楼之一，成为一个家庭所拥有的周末临时住所，他们经常到郊区去休闲活动。在这个 1575 平方英尺（合 146m²）的三卧室单元的入口处有一个绿色的金属塔楼，使主人想起自己所喜爱的在海上的航行浮标。这座小塔有两个作用，另一个作用是它还是一面防火墙，这是按伦敦防火规范的要求而设置的。在塔的背后是儿童卧室，设有一部梯子可以爬到床铺平台。阁楼的西南端是主起居区，与厨房连在一起。主卧室位于相反的另一端，以获得最大限度的私密性。蓝色的粉末涂料与枫木地板形成鲜明对比，门和厨房采用了一些樱桃木和不锈钢来进行细部处理。

左图：嵌入式储藏室采用了蓝色粉末涂料，与绿色金属塔形成强烈的对比。竖塔的背后是儿童卧室，有梯子通往床铺平台；下面是书桌、座椅和储藏区

LEFT Powder blue paint on the built-in storage units contrasts with a metallic green tower. On the reverse side of the tower is a child's bedroom with a ladder to a bed deck; desk, sitting, and storage areas are below.

对页图：在主起居区，主人尽可能少地设置家具，以突出嵌入式细部构件的美

OPPOSITE In the main living area the owners chose to install minimal furnishings to highlight the beauty of built-in details.

ABOVE Built-in cherry breakfast bar features a dropped ceiling to divide kitchen from main living area.

RIGHT View along the main dividing wall that creates privacy and volume in the loft.

上图：固定式的樱桃木早餐吧台上是一片较低矮的吊顶，将厨房与主起居区区分开来

右图：沿主要分隔墙处的内景，营造出一种更为强烈的私密性和楼阁中的体量

ABOVE View along the main access of the loft.

RIGHT Push-pull stainless-steel door handles were designed especially for this project.

上图：沿阁楼主通道的内景

右图：不锈钢推拉门的把手是专门为该项目而设计的

TORONTO

Tribeca Loft
TERRELONGE DESIGN

多伦多

特里贝卡阁楼
特里朗奇设计事务所

在加拿大的许多城市里，宽大的纽约式风格的工作室阁楼项目成了时髦，不管是在全新的建筑里还是在商业建筑的改造中都是如此。在这个位于多伦多郊区的阁楼改造区中，开发商邀请德尔·特里朗奇事务所设计了一系列的样板房。大量的细部都是为这些阁楼专门设计的，其中包括埋入地下的浸泡式浴缸。最小的单元面积为680平方英尺（合63.2m²），设有定制的固定家具和9英尺（合2.7m）高的顶棚。由于建筑物中许多原有的要素被隐藏了起来，因而整个平面布置得干净、舒适、令人感到亲切，并且还非常简洁。木制的细部构件、石板楼面、通高的磨砂玻璃门，以及一个温泉浴室，给人一种高雅生活的感受。它同不久前艺术家风格的粗糙仓库所表现出来的歇斯底里有着很大的不同，新的一批阁楼住户们似乎更喜欢享受无所拘束的奢华生活。

上图：主浴室　　ABOVE Master bathroom.

对页图：起居区内景，在通高的磨砂玻璃门的后面是卧室　　OPPOSITE View of sitting area with bedroom behind frosted-glass floor-to-ceiling doors.

ABOVE View of the living area with built-in sofa, slate floors, and painted walls.

上图：起居区里的固定式沙发、石板铺楼面，以及涂料墙面处理

上图：厨房及用餐区以仿石面砖以及定做的橱柜和书架为特色

ABOVE Kitchen-dining area features stone tile and custom cabinetry and bookshelves.

MILAN

Casa di Libri
ROSANNA MONZINI

米兰

卡萨之家图书馆
罗桑那·蒙奇尼事务所

在意大利，将商业建筑改建为居住建筑的情况很少见。在米兰市中心，对一个空间巨大但采光不佳的建筑进行了改造，这是一个普通的带阳台的出租房，在改造后它成了一个阁楼式的住宅，供一个三口之家居住，这家人拥有大量的书籍。起居间的转角处被打开，并被分隔成几个房间，形成室内观景通道，从而改善了采光条件。阁楼式空间的多功能特性在这个住宅中体现了出来，在这个建筑中设置了一个图书馆，住宅的其余部分设置在其周围。在建筑中还增设了铁制楼梯和阳台，作为通往卧室和浴室的通道，让二层可以敞开。两个两层高的内部庭院种满了树木和花草，形成了更多的风景，并将更多的光线引入到建筑之中。

上图：图书馆是这个阁楼式空间中的一个重要部分

ABOVE The library is an important part of the loft-like space.

对页图：两层楼高的庭院里种满树木和花草，并将更多的光线引入到建筑之中

OPPOSITE Double-height courtyards filled with trees and flowers bring light into the building.

上图：图书馆的入口，展示出图书收藏者的梦想

对页图：加建的平台和铁制楼梯使二层完全开敞，从而形成了空间的流动

ABOVE Entrance to the library, a *bibliothèque's* dream.

OPPOSITE The addition of balconies and iron stairways allowed the second floor to be opened for a greater flow of space.

TORONTO

Merchandise Mart Lofts
CECCONI SIMONE

多伦多

玛特商品仓库
赛可尼·西蒙事务所

在多伦多，阁楼设计还处在一个初期阶段，只不过是在1990年代中期才着手进行这方面的工作。开发商们希望能够从这样一个世界性的潮流中获得利益，所以将其注意力转向了一些非常巨大的废弃仓库建筑，将它们改建成住宅。该项目据称是在北美的同类工程中最大的一个，该项目的任务是在以前的西尔斯大楼（Sears Building）的100多万平方英尺（合93000㎡）的空间内营造出舒适的阁楼公寓单元。但是，要想在家庭生活方式上推销这种开放空间的想法却并非易事。因此，样板房的室内规划布置经过了精心的考虑，设计师安娜·西蒙和爱莱茵·赛可尼提出了一些很有创意的设计方案。大部分的阁楼单元都是小而窄，从625到1200平方英尺（合58~111.5㎡）不等，因此，设计师对自然采光问题和空间的实用性及合理的限定做了细致的考虑。

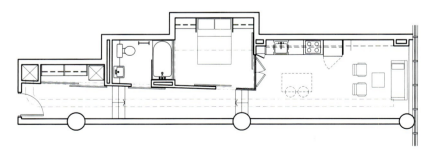

上图：从厨房区看向入口门厅

左图：标准单元平面图

对页图：在这个跃层单元中，开敞楼梯增加了空间的尺度感

ABOVE View from kitchen area down through entrance hallway.

LEFT Typical unit floor plan

OPPOSITE An open stair in this two-level unit adds to the sense of spaciousness.

上图：在这个最小的单元里，高敞的顶棚弥补了活动空间的不足

ABOVE In this, the smallest unit, high ceilings compensate for the narrow footprint.

上图：由于在这个 635 平方英尺（合 58m²）的阁楼中对空间和灵活性要求很高，因此，将不锈钢工作台兼做用餐区

右图：在同一栋建筑中的最大单元里，磨砂玻璃和滑动木门扇根据对私密性的需要而起着一种临时性的功能

ABOVE Stainless steel work table doubles as an eating area where space and flexibility are at a premium in this 625-square-foot (58-square-meter) loft.

RIGHT In a larger unit of the same building, frosted glass and sliding wooden panels serve a temporary function, depending on the need for privacy.

MERCHANDISE MART LOFTS 145

TOKYO

West Shinjuku
FREDERICK FISHER & PARTNERS ARCHITECTS

东京

西新宿住宅

弗德里克·费歇尔联合事务所

西新宿住宅是在东京市内的一个最新的阁楼开发项目，共有三十个住宅单元。该项目重点需要解决空间和自然采光的问题。每个单元都是用混凝土与钢材作为基本的结构，但是在形状和大小上各不相同。建筑师充分利用了材质、色彩和暖色调的木材来丰富建筑效果，并在以灰色为主的环境中脱颖而出。这个大型项目的平面简洁而富有效率，不仅给人一种轻快的感觉，也增加了空间的开敞性，这对于一个在一般居住区内很难找到大空间的城市来说尤为重要。狭窄的过道也被取消了。暴露在外的楼梯、通高楼层的窗户以及夹层等使这些有两层高的阁楼单元拥有良好的视线。宽敞的起居区和有限的外墙面在这个巨大的现代都市中扮演了一个新的角色。

上图：混凝土与钢结构的建筑物为这个在东京的大型阁楼项目提供了可能性

左图：标准层平面图

对页图：住户采用简洁的家具布置来增加空间的宽敞感

ABOVE A concrete and steel building provides the setting for a large-scale loft project in Tokyo.

LEFT Typical floor plans.

OPPOSITE Residents have chosen light furnishings to increase the enjoyment of spaciousness.

上图：夹层空间因拥有俯瞰城市的良好视野而显得明快

对页图：通高的窗户提供了自然采光和一种空间感，这在东京是非常重要的因素

ABOVE Mezzanines are alight with views over the city.

OPPOSITE Floor-to-ceiling windows offer natural light and a sense of space—at a high premium in Tokyo.

SAN FRANCISCO

Bay Loft
BRAYTON & HUGHES DESIGN STUDIO

旧金山

海湾顶楼仓库

布莱顿与休斯设计工作室

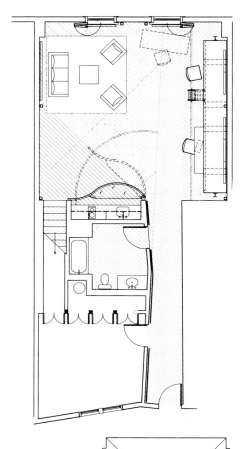

这是在旧金山海湾大桥桥墩附近的一座19世纪的仓库，将其改建成为一个以砖为主要建筑材料的生活与工作用阁楼，其面积为1500平方英尺（合139.4m²）。整洁灵活的空间按照客户的要求形成了充足的采光和工作空间。入口处是放档案的搁架，进去后有一面安装了铰链和滚轮的曲线形枫木墙，墙后是一个小型玻璃台作为厨房使用。当墙打开的时候，厨房就露了出来，而在关上时，它将走道处的墙面镶板沿伸到主要起居和工作区。沿东向墙面的高大窗户朝向海湾，形成了令人愉快的自然采光。在建筑原有的钢梁之外增加了两根新梁，以形成一个规整的网格来悬挂灯具。在楼上设有舒适的卧室和浴室。建筑师把他们所选用的材料品种局限于松木地板、枫木柜子和镶板，以及精轧不锈钢和铝材构件，用它们来把过去和现在联系起来。

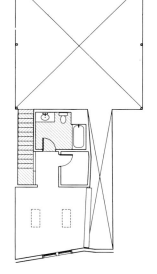

左图： 平面图

对页图： 选择材料的出发点是希望把原来的工业建筑空间与现在联系起来。阁楼里安装了松木地板，枫木做的柜子和镶板，以及精轧不锈钢和铝材构件

LEFT Floor plans.

OPPOSITE The choice of materials was driven by a desire to bridge the past of the former industrial space to the present. The loft features pine floors, maple casework and paneling, and milled-finish stainless-steel and aluminum.

上图： 6英寸（合15.2cm）宽的黄色松木地板沿一条从大厅的一角延伸贯穿整个主厅的缝隙铺设。这一细微的划分对工作区域进行了很好的限定

ABOVE Six-inch (15.2-centimeter) yellow pine laid diagonally along a seam extends from the angle of the hall across the main hall. This subtle division defines the work area.

上图：厨房和通向位于上层的卧室与浴室的楼梯

右图：当安装了铰链的枫木制墙体被打开时的厨房一角

ABOVE View of the kitchen and stairs leading to upper bedrooms and bathrooms.

RIGHT View of the kitchen with the hinged maple wall open.

DIRECTORY
事务所目录

DESIGNERS

Alexander Gorlin Architect
137 Varick Street
New York, NY 10013
tel. 212.229.1199
fax 212.206.3590

Anders Associates
P. O. Box 2710
Midland, MI 48641
tel/fax 517.832.7030

BAK Architecture
Laura Briggs and Paola Iacucci
71 Barrow Street, No. 18
New York, NY 10014
tel/fax 212.255.9867

Belmont Freeman Architects
110 West 40th Street
New York, NY 10018
tel 212.382.3311
fax 212.730.1229

Brayton & Hughes
Design Studio
250 Sutter Street, Suite 650
San Francisco, CA 94108
tel 415.291.8100
fax 415.434.8145

Carpenter/Grodzins Architects
35 East 19th Street
New York, NY 10003
tel. 212.254.2753

Cecconi Simone
663 Queen Street East
Toronto, Ontario
M4M 1G4 Canada
tel 416.462.1445
fax 416.462.2577

Cha & Innerhofer
Architecture + Design
611 Broadway, Suite 540
New York, NY 10012
tel 212.477.6957
fax 212.353.3286

Dean/Wolf Architects
40 Hudson Street
New York, NY 10013
tel/fax 212.732.1887

Frankel + Coleman
727 South Dearborn Street
Chicago, IL 60605
tel 312.697.1620
fax 312.697.1622

Frederick Fisher & Partners
12248 Santa Monica
Boulevard
Los Angeles, CA 90025-2518
tel 310.820.6680
fax 310.820.6118

Galia Solomonoff Projects
249 West 29th Street
New York, NY 10001
tel 212.268.1569
fax 212.631.0379

Hardy Holzman Pfeiffer
Associates
902 Broadway
New York, NY 10010
tel 212.677.6030
fax 212.979.0535

Hut Sachs Studio
55 Crosby Street
New York, NY 10012
tel 212.219.1567
fax 212.219.1677

Kar Ho Architect
117 West 17th Street
New York, NY 10011
tel 212.237.3450

LOT/EK Architecture
Ada Tolla and Giuseppe Lignano
55 Little West 12th Street
New York, NY 10014
tel. 212.255.9326

McDowell + Benedetti
62 Roseberry Avenue
London EC1R 4RR
ENGLAND
tel 171.278.8810
fax 171.278.8844

Moneo Brock Architects
145 Hudson Street
New York, NY 10013
tel 212.625.0308
fax 212.625.0309

Peter Wadley Architects
Shoreditch Studio 44-46
Scrutton Street
London EC2A 4HH
England
tel 171.377.2777
fax 171.377.5439

Rick Mather Architects
123 Camden High Street
London NW1 7JR ENGLAND
tel/fax 171.284.1727

RoTo Architects, Inc.
Michael Rotundi
and Clark Stevens
600 Moulton Avenue
Los Angeles, CA 90031
tel 323.226.1112
fax 323.226.1105

Scott Marble • Karen
Fairbanks
Architects
66 West Broadway, #600
New York, NY 10007
tel 212.233.0653
fax 212.233.0654

Terrelonge Design
477 Richmond Street West
Toronto, Ontario
M.SV3G7 CANADA
tel 416.564.5342

Tod Williams, Billie Tsien and
Associates
222 Central Park South
New York, NY 10019
te 212.582.2385
fax 212.245.1984

Tow Studios
Peter Tow
260 5th Avenue, Suite 1206
New York, NY 10001
tel 212.576.1807

Vicente Wolf Associates, Inc.
333 West 39th Street
New York, NY 10018
tel 212.465.0590
fax 212.465.0639

DIRECTORY

摄影工作室目录

PHOTOGRAPHERS

Bjorg Arnarsdottir
Bjorg Photography
517 Avenue of the Americas
New York, NY 10011
tel 212.255.5258

Arcaid
The Factory
2 Acre Road
Kingston on Thames
Surrey KT2 6EF Great Britain
tel 44.181.546.4352
fax 44.181.541.5230
Dennis Gilbert
Alberto Piovano

Rico Bella
19 Lysander Court
Toronto, Ontario
MIV 3R2
tel 416.321.1364

Benny Chan
Fotoworks
824 17th Street, #5
Santa Monica, CA 90403
tel 310.449.0026
fax 310.264.9777

Billy Cunningham
140 7th Avenue
New York, NY 10011
tel 212.929.6313
fax 212.929.6318

Esto Photographics, Inc.
222 Valley Place
Mamaroneck, NY 10543
tel 914.698.4060
fax 914.698.1033
Peter Aaron
Jeff Goldberg

Hedrich Blessing
Photographers
11 West Illinois Street
Chicago, Illinois 60610
tel 312.321.1151
fax 312.321.1165
Marco Lorenzetti

Eduard Hueber
Arch Photo, Inc.
51 White Street, #5S
New York, NY 10013
tel. 212.941.9294
fax 212.941.9317

David Joseph
Snaps
523 Broadway, #5
New York, NY 10012
tel 212.226.3535
fax 212.334.9155

Chun Y. Lai Photography
119 West 23rd Street,
Studio 905
New York, NY 10011
tel. 212.645.2385
fax 212.691.1668

Norman McGrath
Photography
164 West 79th Street
New York, NY 10024
tel 212.799.6422
fax 212.799.1285

James Mitchell
62 Roseberry Avenue
London EC1R 4RR ENG-
LAND
tel 171.278.8810

Michael Moran Photography
371 Broadway, 2nd floor
New York, NY 10013
tel 212.334.4543
fax 212.334.3854

Peter Paige Associates, Inc.
269 Parkside Road
Harrington Park, NJ 07640
tel 201.767.3150
fax 201.767.9263

Tim Soar
62 Roseberry Avenue
London EC1R 4RR ENGLAND
tel 171.278.8810

John Sutton Photography
8 Main Street
Pt. San Quentin, CA 94964
tel. 415.258.8100
fax. 415.258.8167

Joy von Tiedemann
24 Tennis Crescent
Toronto, Ontario M4K IJ3
Canada
tel 416.465.0843
fax 416.465.3334

Paul Warchol
224 Centre Street, 5th Floor
New York, NY 10013
tel 212.431.3461
fax 212.274.1953

Christopher Wesnofske
Photography
280 Park Avenue South, #16C
New York, NY 10010
tel/fax 212.473.0993

PHOTOGRAPHY CREDITS
照片来源

NEW YORK CITY

Urban Interface Loft, Dean/Wolf Architects, pp.14–17
Photographs by Peter Aaron/ESTO

The Ultimate Loft, Hardy Holzman Pfeiffer Associates, pp. 18–23
Photographs by Norman McGrath

34th Street Loft, LOT/EK Architecture, pp. 24–27
Photographs by Paul Warchol

Quandt Loft, Tod Williams, Billie Tsien and Associates, pp. 28–31
Photographs by Peter Paige

Family Living Loft, Scott Marble • Karen Fairbanks, Architects, pp. 32–37
Photographs by Eduard Hueber, Arch Photo, Inc.

Tribeca Loft, Tow Studios, pp. 38–41
Photographs by Bjorg Arnarsdottir

Renaud Loft, Cha & Innerhoffer Architecture + Design, pp. 42–45
Photographs by David Joseph

Fifth Avenue Loft, Scott Marble, Karen Fairbanks, Architects, pp. 46–49
Photographs by Eduard Hueber, Arch Photo, Inc.

Divney Residence, Hut Sachs Studio, pp. 50–53
Photographs by Jeff Goldberg/ESTO

Hudson Loft, Alexander Gorlin Architect, pp. 54–57
Photographs by Peter Aaron/ESTO, and Billy Cunningham

Wei Loft, BAK Architecture, pp. 58–61
Photographs by Eduard Hueber, Arch Photo, Inc.

City Loft, Vicente Wolf Associates, Inc., pp. 62–65
Photographs by Vicente Wolf

Chelsea Loft, Kar Ho Architect, pp. 66–71
Photographs by Bjorg Arnarsdottir

Rosenberg Residence and Studio, Belmont Freeman Architects, pp. 72–75
Photographs by Christopher Wesnofske

Live/Work Dualities, Dean/Wolf Architects, pp. 76–79
Photographs by Eduard Hueber, Arch Photo, Inc.

Artist's Studio, Galia Solomonoff Projects, pp. 80–83
Photographs by David Joseph

Tribeca Home and Studio, Moneo Brock Architects, pp. 84–87
Photographs by Michael Moran

GOING GLOBAL

Los Angeles
Carlson-Reges Residence, RoTo Architects, Inc., pp. 90–93
Photographs by Benny Chan, Fotoworks

Chicago
Concrete Loft, Frankel + Coleman, pp. 94–99
Photographs by Marco Lorenzetti for Hedrich Blessing

Antwerp
Coolen House, Kris Mys Architect, pp. 100–103
Photographs by Alberto Piovano for Arcaid

London
Oliver's Wharf, McDowell + Benedetti, pp. 104–109
Photographs by Tim Soar

Avoca, Pennsylvania
O'Malley Residence, Carpenter/Grodzins, Architects,
pp. 110–113
Photographs by Chun Y. Lai

London
New River Head, McDowell + Benedetti, pp. 114–117
Photographs by James Mitchell

Jersey City
Pavonia Loft, Anders Associates, pp. 118–123
Photographs by Otto Baitz and Simo Neri

London
Neal's Yard, Rick Mather Architects, pp. 124–129
Photographs by Dennis Gilbert for Arcaid

London
Chiswick Green Studios, Peter Wadley Architects, pp. 130–133
Photographs by Peter Wadley

Toronto
Tribeca Ontario, Terrelonge Design, pp. 134–137
Photographs by Rico Bella

Milan
Casa di Libri, Rosanna Monzini, pp. 138–141
Photographs by Alberto Piovano for Arcaid

Toronto
Merchandise Mart Lofts, Cecconi Simone, pp. 142–145
Photographs by Joy von Tiedemann

Tokyo
West Shinjuku, Frederick Fisher & Partners, pp. 146–149
Photographs by Motoi Niki and Frederick Fisher

San Francisco
Bay Loft, Brayton & Hughes Design Studio, pp. 150–153
Photographs by John Sutton

译 者 简 介

重庆仁豪景观设计有限公司前身为1997年成立的仁豪设计工作室，公司人员近30人。在近四年的时间内，已先后完成数十项规划、建筑及城市景观设计，并得到社会好评。公司正式成立于2001年3月，翌年年初即获重庆市2001年度规划设计奖及小城镇风貌设计奖两项奖励。

公司本着"思想创造财富"的理念，力图逐步发展成为一个跨越城市规划、建筑设计、城市设计、景观设计的服务提供商和新兴的高科技咨询专业公司，在致力于创作大陆本土设计精品的同时，积极联合国内外优秀的合作者和介绍国外设计新思潮（已翻译《建筑与艺术》一书），是仁豪公司始终不渝的目标。